LABORATORY MANUAL

Sandra Chimon Rogers

Calumet College of St. Joseph

With Contributions by
Stephanie Dillon, *Florida State University*
Barbara Mowery, *York College of Pennsylvania*

CHEMISTRY

McMURRY • FAY • ROBINSON

SEVENTH EDITION

PEARSON

Boston Columbus Indianapolis New York San Francisco Upper Saddle River
Amsterdam Cape Town Dubai London Madrid Milan Munich Paris Montreal Toronto
Delhi Mexico City Sao Paulo Sydney Hong Kong Seoul Singapore Taipei Tokyo

LABORATORY MANUAL
CHEMISTRY

SEVENTH EDITION

SANDRA CHIMON ROGERS

WITH CONTRIBUTIONS BY STEPHANIE DILLON AND BARBARA MOWERY

This comprehensive laboratory manual features 29 experiments with a focus on real-world applications. These experiments were written specifically to correspond with the Seventh Edition of *CHEMISTRY* by McMurry/Fay/Robinson. Each experiment explores one or more topics, covered within a chapter in the textbook, with the goal of helping students understand the underlying concepts covered in the lecture course.

The format of each experiment is devised to replicate a typical research experience.

EACH EXPERIMENT CONTAINS:

- A pre-laboratory report that should be completed before starting each experiment.
- An introduction, including a discussion of what the experiment entails, why the experiment is important to students as chemists, and where the concept or technique learned can be used in the real world.
- A background section explaining concepts that each student is expected to master for a full understanding of the experimental results. This section will often contain any example calculations needed to complete the laboratory report.
- Step-by-step procedures.
- A report section separated into the traditional sections of a scientific journal article. This section also contains details about what is expected in the report, as well as thought-provoking questions that should be answered within the report's conclusion.

ISBN-13: 978-0-13-388662-7
ISBN-10: 0-13-388662-X

EAN

9 780133 886627

90000

PEARSON

www.pearsonhighered.com

Table of Contents

To the Student:

It is hard to imagine this, but chemistry is everywhere, chemistry can be fun, and chemistry can be challenging. However, do not let the latter be the reason you do not take the class and enjoy the lab. I know that you are probably thinking to yourself, "I hate chemistry" or "It's too hard." Or maybe you're even asking the questions "Why are we learning this stuff, and when am I ever going to apply this?" Well, with over 15 years of teaching experience, including the time I was an undergraduate, I have worked with all types of students, from the traditional to nontraditional, to those who have had a year of high school chemistry to those who have never taken a math course. And what I can tell you is that once you work in the lab setting, things change. You have fun trying new things. You have fun seeing real-life examples, and in the entire process, you are applying what you learned in lecture to an actual situation where it is used.

Chemistry lectures are only part of the equation to successfully learning the material. Some people, such as myself, are more hands on. We can talk about concentrations for days, and if I don't get to actually see it in the lab, then how do I know that what I am being taught is factual? That's the beauty of these labs. I can see when and how the lecture materials are applied.

Not only are there scientific applications in this lab manual, also included are real-life applications. Each experiment discusses what it entails along with why it is important and where the concept and/or technique will be applied again. The main reason being is for understanding. However, these concepts and techniques will be used over and over again by the serious science student. The labs are designed to replicate a typical research-style experience. What I mean by this is that the experiments in this manual are investigative in nature, they always require some preparation on the part of the student prior to experimentation, and they are always completed by production of a written report in traditional scientific journal format.

As the author of this manual, I hope that you have fun doing science, have fun with chemistry, and at the same time, with the assistance of this lab manual, understand the how's and why's of the experiments. As with any written work, there is always room for improvement and the potential for typos and errors. Please feel free to contact me if you find any such errors in the text or if you see a place for improvement.

Sandra Chimon Rogers, PhD
Department Chair for the Sciences, Mathematics and Behavioral Sciences
Calumet College of St. Joseph

Laboratory Safety

Safety is of paramount importance in any laboratory. The laboratory safety rules here should be considered as a starting point for safe laboratory practice. Additional Safety Notes are included in each of the experiments in this manual. In addition, your instructor will give specific guidelines on the safe conduct of each experiment at the beginning of the laboratory period. Non-attendance during this safety briefing can result in students being barred from participation in the laboratory experiment.

The following comments on the general rules as listed may help to clarify any questions which arise. Please inquire of your instructor if there are further questions.

1. **Working without supervision is forbidden.**

   A student should never stay in the laboratory alone or work without an instructor present.

2. **Performing unauthorized experiments or any experiment at unauthorized times is forbidden.**

   Unanticipated results to an unauthorized or altered experiment can be quite hazardous. Materials should never be taken from the laboratory. Horseplay and pranks are never acceptable in the laboratory.

3. **Approved safety eye wear must be worn at all times.**

   Certain solvents in use during some experiments can have adverse effects on soft contacts and students should take appropriate precautions following those labs. Consult your instructor about the policy for wearing contact lenses in your laboratory.

4. **Loose hair and clothing must be restrained.**

   Long hair and loose clothing that can get caught on glassware, catch on fire or otherwise be a hazard in lab should be pulled back.

5. **Appropriate attire must be worn in lab.**

   Shoes which cover the entire foot, and clothing which covers the legs and torso are required. Clothing items such as halter and midriff tops, as well as shorts, skirts and capri pants, are inappropriate for laboratory work, as are sandals and flip-flops. Clothing worn should be capable of providing a barrier between your skin and the chemicals you are working with. Coverage of the lower legs and feet are necessary in case of dropped glassware.

6. **Eating, drinking, smoking, chewing tobacco, dell phone use and applying cosmetics are forbidden in the laboratory.**

   These activities should take place only outside the laboratory doors.

7. **All accidents and breakage must be reported to an instructor.**

   Any accident requires attention to clean up. Alert your neighbors immediately if there is a spill. Be sure to warn your classmates of the nature and extent of the spill before you leave the area.

8. **Pipetting by mouth suction is forbidden.**

   Pipetting by mouth suction can lead to accidental ingestion of chemicals and is an unnecessary risk. Pipette bulbs and/or dial-up pipetters will be provided when necessary.

9. **When needed, gloves must be worn.**

   Gloves should be of a material and thickness appropriate for the reagents being used. However, gloves provide only a temporary layer of protection against chemicals on your skin and may be permeable to some chemical reagents, without visible deterioration. If your gloves come in contact with a chemical reagent, remove them, wash your hands, and get a new pair immediately. Hands must be washed just before leaving the laboratory. Students should remove gloves prior to leaving the lab. Any student with a latex allergy should alert their instructor so other arrangements can be made.

10. **All laboratory workers must know the location and proper use of all laboratory safety equipment,** including eyewash, safety shower, fire extinguishers, and telephone.

    Your instructor will show you the location of and proper use of laboratory safety equipment.

11. **All laboratory workers must know how to safely evacuate the laboratory in the event of an emergency.**

    You should note all possible exits from your laboratory. Emergency classroom evacuation requires that the class reassemble away from the building to call roll and check for complete evacuation. Check with your instructor to confirm where the class should reassemble.

ADDITIONAL SAFETY NOTES

The following items were not covered in the general safety rules detailed above, but should be noted in connection with general chemistry laboratories.

1. Items such as book bags, backpacks, purses and coats should be put out of the way off the floor and lab bench.

2. Cell phones should be turned off before entering the laboratory.

3. Read labels and dispose of any waste in an appropriate waste container. Ask questions if you are not sure what to do.

4. Clean up in the laboratory is essential. Make sure you wash and put away all equipment before you leave the laboratory.

6. Inform your instructor at the beginning of the term if you have any special medical conditions that may need attention during the laboratory, such as known allergies or if you are pregnant or suffer from seizures.

Laboratory Notebooks and Reports

Accurate records and reporting of experimental results and conclusions are an indispensable part of any scientific work. A laboratory notebook should be written and organized so that it can be understood by anyone conversant with the subject. All data taken in the laboratory should be recorded in the notebook, NEVER write something on a loose sheet or scrap of paper to be "recorded" later. On the other hand, laboratory reports allow time for thoughtful analysis of the experiment performed and the data collected. Hence the two are similar to taking notes prior to writing a term paper; the first is an immediate record, the second is what you present as your original work. Examples of the two follow at the end of this discussion.

LAB NOTEBOOKS

1. A bound notebook with numbered pages is the standard style of notebook used for research. Carbonless copy notebooks are often preferable as they allow your instructor to keep a copy of your data. Your instructor will tell you which type of laboratory notebook is required for your class. Make sure you read the rules on how to keep a notebook before writing in it for the first time. Some general rules are listed below:

   Notebook Organization:

   The **Title Page,** included with the **Table of Contents,** is the front page of the notebook. Make sure you fill in the following information as soon as possible so that your notebook can be returned to you if misplaced:

   Name
   Local Address
   Local Phone
   E-mail Address
   Lab course section, time and place of meeting
   Instructor or Teaching Assistant's name

   The **Table of Contents** should have the date, title and beginning page number of each experiment.

   Each **Experiment** should be numbered, and have the title, date, and your name on each page, as well as your class and section number.

   If it is not pre-numbered, write the page number on every page of the notebook before using the notebook.

2. Before coming to lab, make sure to read the experiment to be done. Make notes in your notebook and work out a tentative procedure and data tables to be filled in.

3. Never remove the original numbered pages.

4. Write all entries into your notebook in ink.

5. If an error is made in recording data, draw a single line through the mistake and correct it nearby. If a substantial portion of a page is to be discarded, cross the material out with an X. You should never "obliterate" data. This can be interpreted as the scientist having something to hide and all science should be available for review (even the mistakes).

6. All entries should be appropriately labeled with units and in the correct number of significant figures when appropriate. If there are calculations which need to be worked out in class, do them in the notebook.

7. Sign and date each page. Be sure that the name of your lab partner(s) is recorded if applicable.

8. Have your notebook initialed by the instructor at the end of each lab period and turn in a copy of your data before you leave, if required.

9. Neatness and some care in taking notes are quite important – notes should be intelligible, but an *immediate record*.

LABORATORY REPORTS

One of the important phases of carrying out any experimental project is the reporting of the results to interested individuals. Rightly or wrongly, in this course as well as in future professional activities, performance will be judged frequently on the basis of written reports. It is therefore very important that students learn to prepare and submit well-organized reports. The following format may help organize the information desired into useful reports:

1. Title.
   Always submit a report with the formal title of the experiment. Include the names of lab partners if there are any, as well as the date, your instructor's name and the lab section number.

2. Purpose or Abstract.
   Give a brief but complete statement describing the major goals of the experiment. This should not simply be a regurgitation of what is written in the lab manual but your own thoughts on what will constitute a successful conclusion to the experiment being performed.

3. Procedure.
   In the formal report of an experiment *you* have designed, this section describes how you went about doing the experiment. Since in most cases for this lab, the procedure is already described, students need only cite the reference to the laboratory manual. Remember to note in this section any changes or deviations made to the procedure referenced. (Complete sentences are best.)

4. *Data.*

Present summarized experimental data carefully and neatly. This need not be as complete a data table as in the lab notebook, but should contain all the pertinent data used in calculations. Headings should be enough that they stand alone without reference to the manual. It may be convenient to include calculated values and results in this section for clarity and ease of reference.

5. *Calculations.*

Show all equations and calculations (neatly organized and documented) used to arrive at final results, and show the final results clearly. For repetitious calculations, you may show the general formula and one example. This section should also contain any graphs and further tables. Units and significant figures must be used correctly in all calculations.

6. *Conclusions.*

Prepare a brief conclusion in which the results of the experiment are discussed. Was the purpose stated at the beginning achieved? If not, why not? What would you do differently if you were to repeat the experiment in order to get a better result? Include in this section a discussion of major possible errors and their projected impact on the experiment. Also in this section, answer any questions or exercises included as part of the experiment.

NOTE TO STUDENTS: Constructive written criticism of the experiments and/or their reports is welcome and will be given serious consideration.

Some Common Laboratory Glassware and Tools

Glassware or Tools	Common Use	Glassware or Tools	Common Use
Beaker	Beakers are used to contain volumes of liquids, carry out reactions and heat solutions.. They are not volumetric and their accurate only to +/−5% of their graduation. They should not be used to measure volumes of liquids.	Funnel	Funnels are used in chemistry for the same purposes that they are used in everyday life; to prevent spillage.
Buret & Buret Stand	A buret is used to deliver solution in precisely-measured, variable volumes. Burets are used primarily for titration, to deliver one reactant until the precise end point of the reaction is reached. Burets are volumetric, and therefore are used to make accurate measurements.	**Graduated Cylinder**	Graduated cylinders are volumetric measuring devices designed to measure and deliver accurate volumes of liquids. When measuring liquids in a graduated cylinder, you always measure from the center of the curvature in the liquid called the meniscus. Some liquids, like mercury, have a convex meniscus. Most liquids, however, will have a concave meniscus and the volume in the graduated cylinder should be read from the bottom center of this meniscus. (See below)

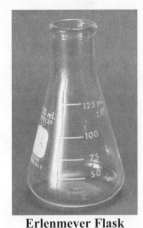 **Erlenmeyer Flask**	Erlenmeyer flasks are used for mixing, transporting, reacting, and filtration but not for accurate measurements. The small neck of the flask prevents liquids from escaping. The volumes stamped on the sides are approximate and only accurate to within about +/−5%. Erlenmeyer flasks should not be used for measuring volumes.	 **Striker**	Strikers are used to start Bunsen burners. The striker itself is constructed with a rough surface positioned opposite to a piece of flint. When the arm of the striker containing the flint is pushed back and forth over the rough surface sparks are produced. These sparks, when created in the presence of a flammable gas such as propane, will start a fire or for our purposes a Bunsen burner.
 **Mohr Pipette**	A Mohr or transfer pipette is used to measure small amounts of solution very accurately. Common pipette volumes range from 0.1 mL to 10 mL.	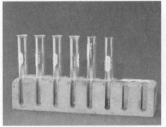 **Test Tubes & Test Tube Rack**	A Test tube is non-volumetric cylindrical container used to observe small volumes of reactants or carry out small size reactions.
 **Watch Glass**	A watch glass is a round, shallow concave piece of glass that can be used to hold liquid or solids for evaporation, to dry chemicals, to weigh chemicals in lieu of a weighboat, or it can be used as a lid for a beaker to prevent dirt from getting into a solution. Because it is not an airtight seal, gases can still be exchanged.	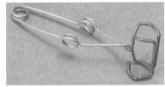 **Test Tube Clamp**	The clamp to the left is a test tube clamp. Clamps are used to allow a person to be "hands-free" during the progress of a reaction. Clamps such as the one shown are used to handle test tubes that are hot or to prevent slippage.
 **Pipette Bulb**	Rubber bulbs are used in conjunction with pipettes to transfer liquids from place to place.	 **Spatula**	Spatulas are used to transport and distribute dry chemical compounds. Spatulas are used most often when weighing out chemicals on a balance because they allow you to transfer very small quantities of the chemical at a time.

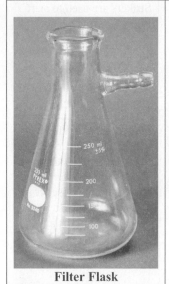

Filter Flask

A filter or side arm flask is an Erlenmeyer flask made of slightly thicker glass to withstand high pressure. The side-arm is conical and used to attach a vacuum hose. A Buchner funnel is placed into the neck of the flask to allow pressurized filtration of solutions.

Volumetric Flask

Volumetric glassware or volumetric flasks are containers that have been calibrated to a specific volume. These flasks, unlike Erlenmeyer flasks or beakers, are marked with lines that are calibrated to a specific volume of liquid. A volumetric flask should be used whenever an accurate concentration of solution is required

Buchner Funnel

A Buchner funnel is a suction or vacuum filtration funnel that is used in conjunction with a filter flask to separate particulates from solution. Filter papers of varying pore sizes are used to specify the size of the particulate retained.

Desiccator

The desiccator is used to store dried samples in a dry atmosphere.

Water Bottle

A wash bottle is generally filled with distilled or de-ionized (DI) water and used to make solutions and rinse glassware. The DI water should be kept fresh as exposure to air will eventually introduce CO_2 and increase the acidity. The wash bottle can be used for any liquid so always label the bottle appropriately.

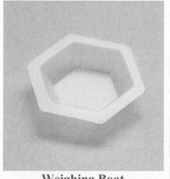

Weighing Boat

Weigh boats are disposable plastic containers used to prevent reagents from contacting the balance pan when they are being weighed.

Experiment 1A
Measurement and Expression of Experimental Data

Purpose

This first laboratory is a review of the mathematics, statistics, and graphing techniques that you will be expected to know and use during your general chemistry courses. The expression and analysis of experimental data is the focus of any science-related field. The handling of this data, its expression, and the interpretation of its meaning is the text of most scientific journals. We, as scientists, devise experiments to collect data that will give us the answer to a specific problem or lend support to a hypothesis. Most often, this data is numerical in form and must be dealt with by a combination of mathematical, graphical, and statistical methods to allow for a reasonable and, more importantly, reproducible result.

Part A of this lab exercise will demonstrate the proper way to manipulate and report experimental data. It includes a basic review of some mathematical manipulation of data as well as the basics of graphing. Part B will cover a review of the statistical methods used to evaluate experimental data and will explore more sophisticated graphing methods, including spreadsheet programs like Excel®. Both sections should serve as a reference for most of the experimental portion of this general chemistry course and any that follow.

Background

Measurements are a way to describe observations with a reference point. To relay the points of reference and measurements to anyone in the world, a set of agreed upon standards has evolved. The current system, established in 1960, is called the International System of Units (abbreviated SI) and has seven basic units from which all others can be derived. They are as follows:

Basic SI Unit	Abbreviation	Measures
Meter	m	Length
Kilogram	kg	Mass
Second	s	Time
Kelvin	K	Temperature
Mole	mol	Amount of substance
Ampere	amp	Power
Candela	cd	Luminescent intensity

Some of these measurement units are quite familiar, others less so. The advantage of this system, as of the metric system which preceded it, is that through the use of various prefixes, powers of 10, and exponential notation, the units can be adjusted to fit almost any sensitivity of measurement, from the mass of the earth ($\sim 5.98 \times 10^{24}$ kg) to the mass of a helium atom (6.644×10^{-27} kg).

For work in the laboratory, we will continue to use some of the older and more familiar terms for units such as liters for volume; atmospheres, torr, or mm Hg for pressure; and Celsius for temperature. These are easily converted to SI units when necessary. We must also be familiar with other units that have been used in the past, including the English system of measures. Although SI and metric units are easily convertible using powers of 10, conversions between English and SI or metric units are often inexact, depending upon the accuracy of the conversion factor between the two systems. Several useful tables are found in the appendices of this lab manual, including tables of conversion factors, physical constants, and metric prefixes.

Measurement

When measurements are taken, there is always some degree of uncertainty. The size and type of uncertainty (and therefore possible errors) depend upon the care with which a measurement is made. These errors typically take two forms: errors arising from the imperfections of the instrument used to make the measurement (miscalibration) and errors arising from the skill or technique of the experimenter. In the first case, the errors are systematic (all in the same direction). In the second case, the errors will tend to be random (sometimes in one direction, sometimes in another). The two types of errors affect measurements differently. Systematic errors affect accuracy (closeness to the true (actual), value). Random errors affect precision (how well a set of data agrees). In taking measurements in the laboratory, we are concerned with both accuracy and precision. One way to minimize the impact of any error is to take multiple measurements of the same value and use a statistical treatment of the set of values obtained, such as average and/or standard deviation to determine the value. While it is seldom possible to determine accuracy (the "true" value is not often known), the size of the standard deviation does give some indication of the precision of a series of measurements.

Significant Figures

In the chemistry laboratory, a number of different tools and instruments are used to make measurements. Most of these instruments are very finely machined or calibrated to produce accurate and precise values. For instance, which do you think would produce a better value for the length of a room, a measurement in notebook lengths, or a measurement produced with the use of a small ruler? No matter how well calibrated a given instrument may be, there is still some degree of uncertainty in the measurement made. Look at the smallest delineating mark on a metric ruler.

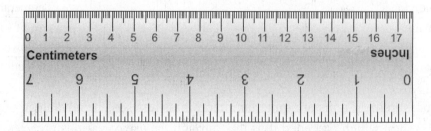

The numbered marks are centimeters and the smallest mark is a millimeter (0.001 m). If an object is measured, it may fall between the 14.1 cm mark and the 14.2 cm mark. An estimate of the length is then made in the next decimal place (one place past the mark(s)), and the object's length is determined to be 14.13 cm. In this example, the uncertainty is ± 0.01 cm because we have estimated the final decimal place. Although the last digit is not certain, it is still reasonably reliable and should be reported. Any digits that are considered reliable are important for calculations involving a measurement, and therefore, these digits are known as **significant digits** or **significant figures**. In using any measuring device, an estimate of one decimal place past the marks should be made; thus, the uncertainty in any measurement is in the last digit. On some instruments, such as the barometer, there is a built-in estimating device called a Vernier scale, which allows the final decimal to be estimated more quickly. In reading various digital displays, the last digit is also assumed to be an estimate, unless the instrument states a different uncertainty. Often, very sensitive devices will state explicitly what the error in the measurement is calibrated to be.

It stands to reason that any time a series of measured values is combined mathematically, the value that is least precisely known determines the precision of the whole set. The following set of rules will help to clarify how to determine the number of significant digits in a measurement and how any combination of measured values should be rounded to show an appropriate number of significant figures in the end result.

Rules for Determining Significant Figures

1) All nonzero digits are significant regardless of their position relative to a decimal point.

    Examples: 123 has three significant figures.
              669988 has six significant figures.
              7 has one significant figure.

2) Any zeros surrounded by nonzero digits, called captive zeros, are significant.

    Examples: 1.101 has four significant figures.
              750.25 has five significant figures.
              303 has three significant figures.

3) Zeros to the left of all nonzero digits (leading zeros) are not considered significant because they serve only as decimal place holders.

    Examples: 0.08206 has four significant figures (the 8, 2, and 6 are significant by
              Rule 1; the 0 between the 2 and 6 is significant by Rule 2).
              0.0003 has only one significant figure because the zeros are all leading zeros.

4) Zeros to the right of the last nonzero digit (trailing zeros) may or may not be considered significant.

    a. If there is a decimal point to the left of any of this type of zero, then they are to be considered significant.

        Examples: 0.03750 has four significant figures (the three nonzero digits plus the trailing zero beside the 5).

                          69.0 has three significant figures.

                          10.000 has five significant figures.

    b. If the decimal point is to the right of this type of zero, then it is necessary to know what the number represents to determine the amount of significant figures.

        Example: The number 100 could have one or three significant figures depending on its context. If it represented an approximation of the volume of water in a glass (about 100 mL), then it would only be considered to have one significant figure. However, if the water had been measured in a graduated cylinder, then all three digits would be significant. In this example, the decimal point that is expressed suggests that the zeroes are meant to be significant.

                    Often, this kind of guessing can be avoided if the measurement is recorded in scientific notation. For instance, in the above example, the number could be written as 1.00×10^2 mL, indicating that there are three significant digits rather than one in the number.

Rounding Numbers

The number of significant figures in a measurement becomes very important when doing calculations with collected data. For example, the density of an object should not be reported to four significant figures if its weight or volume was recorded to only three significant figures. Similarly, the density should not be reported to only two significant figures because useful information would be discarded. **Generally, the calculated result is only as reliable as the least precisely measured value used in the calculation, that is, the number with the fewest number of significant figures.** (In the examples below, all numbers are rounded to the hundredth place for simplicity. The least number of significant figures present in the calculation would determine the actual number to be reported.)

Rules for Rounding

1) If the first digit being dropped is less than 5, then the previous digit is not changed.

    Examples: 1.234 rounds to 1.23.

                 0.091 rounds to 0.09.

2) If the first digit being dropped is more than 5, then the previous digit is increased by 1.

   Examples: 2.3154 rounds to 2.32.
              78.987 rounds to 78.99.

3) If the digit to be rounded is exactly 5 (or 5 followed by zeros), then the number should be rounded to be even.

   Examples: 3.72500 rounds to 3.72.
              0.975 rounds to 0.98.

   This way of rounding is endorsed by the American Chemical Society.

Rounding during Addition and Subtraction

Rounding in addition and subtraction calculations is determined by the least precisely measured value, not just the count of significant figures. The number of digits after the decimal point determines how many significant figures the answer should have.

Examples:

$$
\begin{array}{r|l}
45.5 & 609 \\
0.9 & 75 \\
34.9 & \\
+56.4 & 43 \\
\hline
137.8 & 659
\end{array}
$$

This answer should be rounded to 137.9. **34.9**, the least precisely known value, has only one decimal place; therefore, the answer should be reported to only one digit after the decimal place.

Note that addition or subtraction occurs first and rounding follows as the final step.

Rounding during Multiplication and Division

As stated earlier, a calculated result can only be as reliable as its least precisely measured value. In multiplication and division, this means that the result of a calculation should have the same number of significant figures as the measurement with the fewest significant figures. If this is not done, the precision of the calculated result is overestimated.

Examples: $108.14 \times 0.0015 = 0.16221$ is rounded to **0.16** because 0.0015 only has two significant figures.

$457.2 / 625 = 0.73152$ is rounded to 0.732 because **625** contains three significant figures.

5

The Role of Conversion Factors in Rounding

Numbers in conversion factors can be either exact or inexact. Exact numbers (numbers with no uncertainty) are used when conversions are made from one set of units in the metric system to another set also in the metric system. The prefixes in the metric system are similar to definitions. For example, *centi* means 100; thus, there are exactly 100 cm in 1 m. Both **100** and **1** in this conversion are exact numbers. Exact numbers are considered to have an infinite number of significant figures. Therefore, when doing calculations, conversion factors do not limit the number of significant figures the answer can have. Conversions from the metric system to the English system (and vice versa) involve inexact numbers. There are approximately 30.5 cm in 1 ft. The **1** ft is the quantity being defined, so it is an exact number. However, **30.5** is a rounded value, which makes it an inexact number. Inexact numbers do affect the number of significant figures that can be present in the answer.

Examples: **Exact:** 355 mL $\times$ (1 L/1000 mL) = 0.355 L

Inexact: 27.9654 cm $\times$ (1 ft/30.5 cm) = 0.916893607 ft, but this is rounded to 0.917 ft because of the inexact number (30.5, 3 significant figures) in the conversion.

Scientific Notation

One way to express very large or very small quantities is to use exponential or scientific notation. Using powers of 10 is a reasonable way to specify the precision of a measurement. Very small numbers have a negative exponent, and very large numbers have a positive exponent. The rules on rounding remain the same. Remember in addition and subtraction to adjust the values to have the same exponent before performing the operation.

Examples: $1760 = 1.760 \times 10^3$, showing 4 significant figures
$0.000000530 = 5.30 \times 10^{-7}$, showing 3 significant figures

Addition/Subtraction
$2.761 \times 10^2 + 1.32 \times 10^1 =$
$2.761 \times 10^2 + 0.132 \times 10^2 = 2.893 \times 10^2$

$1.1941 \times 10^{-2} - 8.62 \times 10^{-3} =$
$11.941 \times 10^{-3} - 8.62 \times 10^{-3} =$
$3.321 \times 10^{-3} = 3.32 \times 10^{-3}$

Multiplication/Division
$(2.761 \times 10^2) \times (1.32 \times 10^1) =$
$3.64452 \times 10^3 = 3.64 \times 10^3$

$(1.1941 \times 10^{-2}) / (8.62 \times 10^{-3}) = 1.38526 \times 10^1 = 1.39 \times 10^1$

Significant Figures Using Logarithms

Logarithms have two parts: (1) The characteristic is the portion of a logarithm to the left of the decimal point and reflects the exponent. (2) The mantissa is the portion to the right of the decimal and reflects the value of the measured quantity.

Example: log 559 = 2.747, **2** is the characteristic, and **747** is the mantissa.

The rule for determining the number of significant figures in a logarithm is that the mantissa should have the same number of significant figures as the measured quantity.

Examples: log 7 = 0.8
log 7.0 = 0.85
log 7.00 = 0.845

Likewise, the same rule applies when the operation is reversed (antilog).

Examples: antilog 0.60 = 4.0
antilog 0.602 = 4.00
antilog 0.6021 = 4.000

Graphing

Using graphs to illustrate a set of data and make predictions is a typical experimental approach. In many experiments, one parameter is varied systematically while observing what changes there are in a second parameter, such as observing the volume of a gas when pressure is increased or decreased. If a linear relationship can be found between two parameters, then predictions are possible. Thus, graphs can be used both to visually inspect a set of data for linear behavior and predict other information. To make the best use of graphical data, a few points on the best way to make them are in order.

1) Use a whole sheet of graph paper for each graph. If the goal is to read information from a graph, then shouldn't the graph be as large as possible?

2) Decide which of the two parameters on the graph is independent (usually this is the one that is varied at regular intervals) and which is dependent (the one that is being observed). The independent variable is plotted on the x-axis, the dependent variable on the y-axis.

3) The best graph of a set of data does not always have the origin, (0, 0), on the graph. Look at the highest and lowest data points of the variables to be plotted. Choose convenient numbers slightly higher and lower than those data points and make them the extremes on the graph. Again, the goal is to spread the data out on the graph as much as possible.

4) Label axes clearly. This should include regularly spaced marks and the units of measurement for each axis. A graph will not be of much use if the axis is not linear.

5) Put a title on the graph that will explain something, not just "Graph of y vs. x." For instance, "Volume of argon gas as a function of pressure at 25 °C."

6) Use a ruler or curve to draw the best line through the plotted points. One way of showing the points is putting a circle around each one. Alternatively, draw error bars showing the precision of the measurement taken.

7) If a graph is not linear, don't try to draw a straight line through the points, draw a curve. To get a straight line, try a mathematical manipulation of one or both of the variables, such as inversion ($1/y$) or logarithm ($\log y$) and do another plot. It may take two or three attempts to find something that works to yield a straight line.

8) Once a straight line is obtained, the slope of the line can be found using two conveniently read points on the line, which should not be previously plotted data points. Remember from algebra that slope, m, is $\Delta y/\Delta x$.

9) Using the equation for a straight line, $y = mx + b$, also from algebra, the y-intercept can be found by substituting any (x, y) point and the slope, m, and solving for b.

Report Contents and Questions

As the first experiment of the term, the exercises presented below are to be used to refresh your memory. None of this experiment is actually done in the laboratory. Therefore, since no data has been taken in the laboratory, none should be recorded in your notebook. Instead, this report should be prepared on notebook paper to be turned in at the next lab class meeting.

The report should consist of a title page, purpose, and the exercises below. Please remember all the rules for constructing good scientific graphs. Show all of your work for full credit.

Exercises

1) Indicate how many significant figures are present in each of the following:
   a. 02.608×10^3
   b. 800.07
   c. 398,004
   d. 3.080×10^{-4}
   e. 0.090

2) Round the following to three significant figures (put in scientific notation where necessary):
   a. 231,431
   b. 0.00389
   c. 0.000 080 320
   d. 3928
   e. 44.522

3) Perform the following mathematical operations:
   a. $(6.49 \times 10^5/2) - 45.38 =$
   b. $5.32 \times 10^{-15} \times 2.447 \times 10^{16} =$
   c. $(3.15 \times 10^{-1} - 9.7 \times 10^{-3}) \times [(4.7 \times 10^{-2})/(2.33 \times 10^{-3})] =$
   d. $1.62 \times 10^{-3} + 8.1 \times 10^{-4} =$
   e. $(0.45 + 167.6)/1.6 \times 10^{-2} =$

4) Express the following using scientific notation:
   a. The Bohr radius: 0.000 000 000 052 917 720 859 m
   b. The speed of light, 299,792,458 m/s
   c. Charge of a proton, 0.000 000 000 000 000 000 160 217 648 7 C
   d. Rydberg constant, 10,973,731 m^{-1}

5) Calculate the following and round to the proper number of significant figures:
   a. Antilog 0.485 =
   b. Ln 60 =
   c. $e^{-1.41} =$
   d. Log $3.975 \times 10^{-2} =$
   e. Antilog 6.86 =
   f. Log 8.74 =

6) Calculate the average and standard deviation for the following set of calibration data: 3.8842, 3.8611, 3.8263, 3.8191, 3.8945, 3.8916, 3.8511, 3.8651

7) Graph the following data and answer the question about the resulting line.

Temperature (K)	Pressure (atm)
273	1.10
283	1.14
293	1.18
303	1.22
313	1.26
323	1.30
333	1.34
343	1.38
353	1.42
363	1.46
373	1.50

Q1.   According to the graph, how would you describe the relationship between temperature and pressure (linear or nonlinear)?

8) Create a graph of the absorbance (optical density) versus molar concentration and answer the questions about the resulting line.

Concentration (M)	Optical Density
0.13	0.19
0.28	0.38
0.43	0.56
0.58	0.72
0.73	0.85
0.88	0.96
1.03	1.06
1.18	1.13
1.33	1.18
1.48	1.21

NOTE: The absorbance, or optical density, of a solution for a particular wavelength of light is given by $\log(Io/I)$, where Io is the incident light on a sample and I is the transmitted light. Beer's law states that the optical density of a substance in solution is directly proportional to its concentration (i.e., it has a linear relationship). Not all substances follow Beer's law, but a calibration graph of optical density versus concentration can still be used to measure concentrations of unknown solutions. Because optical density is a ratio, it does not have any units; the expression *absorbance units* is sometimes used.

Q1. Does the solution in your graph follow Beer's law?

Q2. What is the concentration of a solution that has an optical density of 0.28?

Q3. What is the concentration of a solution that has an optical density of 1.01?

9) Using the same data as above, calculate the logarithm of the volume. Graph the pressure versus the $\log_{10}$ volume and answer the questions about the resulting line.

Q1. If this graph is a straight line, then it is of the form $P = m\mathrm{Log}_{10}V + b$, in which m is the slope and b is the y-intercept. Is the graph a straight line or a curved line (nonlinear variation)?

Q2. If possible, calculate the equation of the line for the graph. Show your work.

Q3. Using the resulting equation from question 2, what would be the volume of a system that had a pressure of 5.4 atm?

10) Robert Boyle used an open-end barometer (similar to that in the figure) for the experiments that led to the discovery of Boyle's law, which describes mathematically the effect of pressure on the volume of a gas. He determined the nature of the law by graphing his data. The law states that PV = constant, where P is the total pressure, which is the sum of $P_{atm} + P_{Hg}$. The pressure can be measured in torr, which corresponds to the height of the mercury column in millimeters.

The following table contains some data in an experiment similar to that of Boyle's, though the open-end manometer is considerably larger. The volume of the gas in the closed end of the tube is measured in milliliters, and the pressure recorded is P_{Hg}, the height of the column, not the total pressure.

Pressure (torr)	Volume (mL)
45	21.7
149	19.1
226	17.6
303	16.3
372	15.3
455	14.2
523	13.5
607	12.6
720	11.7

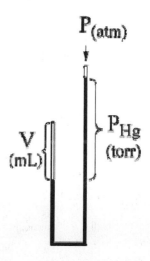

You are to analyze this data by graphing it, discovering that a direct graph (**V vs. P**) gives a curved line. Scientists often try manipulating data to give a straight line, and when the data is regraphed as **P vs. 1/V**, you will see that a straight line is produced, revealing the nature of Boyle's law (that pressure and volume are reciprocally related). Further analysis of this straight line will also reveal the atmospheric pressure at which the measurement is made. Since

$$P_{total}V = \text{constant}$$
$$P_{total} = \text{constant}(1/V)$$
$$P_{atm} + P_{Hg} = \text{constant}(1/V)$$
$$P_{Hg} = \text{constant}(1/V) - P_{atm}$$

the graph is of the form $y = mx + b$, where the slope, m, gives the value of the constant, and the **P**-intercept is the negative of the atmospheric pressure.

Q1. What is the slope of the line (i.e., the constant in the Boyle's law equation)? Give units in your answer.

Q2. What is P_{atm} (i.e., the negative of the **P**-intercept)? Give units in your answer.

11) The following data were used to calibrate a spectrometer. Graph the data and answer the questions regarding the resulting line.

Known Wavelength (nm)	Spectrometer Reading (absorbance units)
388.9	3.47
488.6	4.43
501.6	4.56
587.6	5.38
667.8	6.15
706.5	6.52

Q1. Add a best-fit line and calculate the equation for that line.

Q2. You have collected some more data using the above spectrometer. Use either the graph or the equation of the line to "correct" the data you have collected.

Experimental Spectrometer Reading (absorbance units)	Corrected Wavelength (nm)
5.82	
4.27	
4.00	
6.18	

Experiment 1B
Graphing and Statistical Analysis

Purpose

Scientific results are accumulated in a laboratory setting but are only the backbone to the understanding of science. The ability to express and analyze laboratory results is how scientific fields continue to progress. It is one thing for a person to go into the lab and run an experiment, but those experiments are meaningless unless that data are processed, explained, and defended. This data handling is the body of a scientific text and must give us insights into a specific problem or lend support to a hypothesis.

Most often, this data is numerical in form and must be dealt with by a combination of mathematical, graphical, and statistical methods to allow for reasonable and, more importantly, reproducible results. The handling of data has become easier with the development of computer programs such as Excel®. Spreadsheets are as useful for handling numbers as word processors are for handling words. Spreadsheets are giant data tables that can handle repetitive calculations automatically, requiring much less time and energy than those done by hand. For example, if one piece of data is changed, all calculated results involving that number are immediately and effortlessly recalculated. Excel® also allows for statistical analyses and a variety of graphical representations.

The goal of this lab is to teach you the basics of Excel® and familiarize you with the program, which will make your life easier throughout the semester and in the future. While this lab is going to take effort and what may seem like hours and hours, IT IS going to make your life easier when analyzing data in the future. It is designed to give you practical experience in interpreting the results of raw data from experiments similar to those done throughout the semester and in real life.

One often hears the phrase "A picture is worth a thousand words," and this is true in science. Numerical data can be expressed in tables and paragraphs, but most of the time, a picture would be more helpful. In science, pictures are better known as graphs. Although scientists use graphs as comparison tools, they are more important as a method of determining unknowns. In most cases, known mathematical expressions are rearranged in the form of a straight line, $y = mx + b$ and graphed. Then the slope and y-intercept are determined based on the question of the line.

Graphical representation is commonly used in the calibration of an instrument. To use the calibration curve technique, several standards containing an exactly known concentration of the analyte are prepared. The absorbance of each of analyte standard is then measured using a spectrophotometer. The concentration and the absorbance of the standards are then graphed, with concentration on the x-axis and absorbance on the y-axis. A best-fit line or trend line is made using these points. This is known as the calibration curve. The equation of this line is obtained. Then the absorbance of an unknown solution is measured, and its concentration is calculated.

This technique is used a great deal in testing water for various metals, nonmetals, and ions. For example, the calibration curve technique could be used to test the concentration of nitrate in various water sources. The absorbance of several standards with different known concentrations of nitrate would be found using a Spec-20, and a calibration graph would be made. The absorbance values of the unknown water samples would also be measured. The absorbance values would be used along with the equation of the line to solve for the concentration of nitrate in the water. In this case, the equation $y = mx + b$ is used to solve for x, the nitrate concentration of the water sample, where y is the absorbance value of the water sample, m is the slope of the calibration curve, and b is the y-intercept of the calibration curve.

Another common graphical representation is the extraction of data from the slope of a graph. The slope of the line can provide various information depending on the equation and variables being used for the graph. For example, we know that Gibbs free energy is defined as $\Delta G = -\Delta S(T) + \Delta H$, where ΔS is the entropy, T is temperature in Kelvin, and ΔH is enthalpy. This equation is already in the form of a straight line $y = mx + b$, where ΔG is y, $-\Delta S$ is m, T is x, and ΔH is b. Graphically, ΔG is the y-axis, T is the x-axis, the slope of the line is equal to $-\Delta S$, and the y-intercept is equal to ΔH. This graphical representation allows for the determination of ΔS and ΔH, which are difficult to determine experimentally.

Statistical analysis is important in expressing the reliability of the data being presented. Statistics can be considered an exact treatment of uncertainty. Data values are usually presented with a statistical error. The statistical error is a value written next to the data value stating how much variation in both the positive and negative direction is being experienced. For example, one could write the average concentration as 0.56 +/– 0.01 M, meaning that the average concentration was found to be 0.56 M with a standard error of 0.01 M. If one includes the error, it can be said that the average concentration of the sample is between 0.55 M and 0.57 M. Statistical errors can be caused by multiple sources; common sources are variations between trials and limited accuracy of instrumentation.

Data presented with large statistical errors (standard deviations) are generally not well received by the scientific community. The scientific community wants data with small statistical errors, which corresponds to good precision and accuracy. There are two areas of statistics of great importance to a chemist: the analysis of error in repeated measurements and the analysis of distributions and central tendencies. The first deals with trying to get a feel for the "true" value of a measurement, the second deals with trends in large collections of values or measurements.

The "true" value is obtained when you measure the same thing over and over, and you often end up with slightly different results, despite the fact that you know the values should be the same. For example, it could be measuring the exact miles per gallon your car gets in town, or maybe it's the weight of your pet dog. If you measure your dog's weight to be 105.25 lbs and repeat the measurement five more times, you would expect the dog's weight to be the same, but it's critical to realize that every measurement has an error or uncertainty associated with it. In weighing your dog, the error of uncertainty lies in the accuracy of the scale.

On the other hand, we could count the pennies in a jar and find there to be 125 pennies. The 125 pennies counted is not a measurement, an exact count. But take one of those pennies and measure its density 10 times, and you will likely get 10 slightly different results. These 10 numbers are not random. The density values will all be very similar and, in fact, may be identical to one or two decimal places. They are obviously clustering around a value, the "true" value of the density. There is no way for one to know the "true" value, so we therefore say that we are a certain (90, 95, or 99) percentage confident that the "true" value is within a certain range of the values, the mean +/– the standard deviation.

In addition to analyzing the results of graphical and statistical data, it is important to determine if the results are publishable in a scientific journal such as the *Journal of the American Chemical Society*. The decision to publish is based on the credibility and reproducibility of the data. Accurate data analysis showing the reliability of the results and multiple trials to demonstrate that the results are not just a fluke are both necessary for success. Both graphical and statistical analyses are necessary when determining if the results can be published.

Background

Data Handling

In several laboratory experiments this semester, you will be collecting large quantities of numerical data. To tabulate and express these data, you will need to become familiar with a spreadsheet program. There are several programs of this sort available, but the one we recommend is Excel®. As a Microsoft program, Excel® is available to students on most PCs and is accessible on most university computers. For this reason, this lab recommends the use of and will provide tutorials in Excel®. A tutorial on simple data handling in Excel® is available in Appendix B.

Graphing: The Calibration Graph

Beyond simple data handling, the first technique that will be introduced is how to successfully and concisely express a large quantity of data graphically. No one likes looking at a table full of hundreds of numbers and trying to figure out a particular trend or pattern. What we do like looking at are pictures! And in science, that means graphs. If the same data are presented graphically, the results are much easier to observe as well as to explain.

Types of graphs are as varied as the data they represent, but one of the most common types of graphs used for data analysis, at least in this lab, is the calibration graph. Briefly, this is a graphic representation of standard reference data used to calibrate some variable in an unknown sample. You will use this technique to determine the concentration of phosphate in local water samples using spectrometric methods.

The data table below lists the concentration (mol/L) as well as the corresponding absorbance (in absorbance units, AU) for a set of standard phosphate solutions. If our overall experimental goal is to determine the concentration of phosphate in a water sample obtained from Lake Bradford, we need to analyze our standard data to quantify a trend and then use that trend to determine the concentration of phosphate in our unknown sample. To do this, we generate a scatterplot, like the one shown below, of the dependent variable (absorbance) versus the independent variable (phosphate concentration).

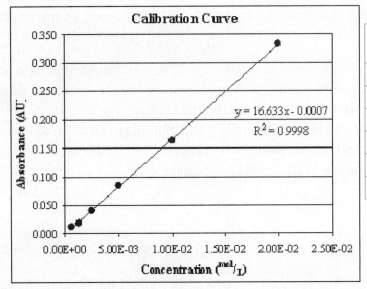

Conc. (mol/L)	Abs. (AU)
2.00×10^{-2}	0.333
1.00×10^{-2}	0.163
5.00×10^{-3}	0.084
2.50×10^{-3}	0.041
1.25×10^{-3}	0.019
6.25×10^{-4}	0.011

Once the scatterplot has been generated, a best-fit line can be added to see if our data can be described as linear, logarithmic, polynomial, or exponential. This process is called linear regression. Remember these different possibilities because not all data has a direct linear relationship. For the example above, the plotting of data results in a linear graph, and we can identify the values of m and b from the $y = mx + b$ form. The actual linear fit equation is shown in the upper right corner of the graph, as well as a corresponding R^2 value. The R^2 value is a statistical measure of how well the best-fit line fit the data. Specifically, the closer the R^2 value is to 1, the better the fit.

Now that we have an equation that describes the trend in the relationship between phosphate concentration and spectrophotometric absorbance, we can use it to determine the concentration of phosphate in our unknown. After obtaining the absorbance of our sample experimentally, we can quickly calculate the corresponding phosphate concentration.

Example: Determining the Phosphate Concentration in an Unknown Sample of Water

According to the best-fit line equation, $y = 16.663x - 0.0007$,

$$\textbf{absorbance} = \textbf{16.663[PO}_4^{3-}\textbf{]} - \textbf{0.0007}$$

We determine this equation simply by substituting in the true values of the independent (x) variable and the dependent (y) variable.

In our example, we have observed that our unknown sample has an absorbance value of **0.126**. If we substitute this value into our equation and solve for $[PO_4^{3-}]$, we will have determined the concentration of $[PO_4^{3-}]$ in our unknown:

$$0.126 = 16.663[PO_4^{3-}] - 0.0007$$

$$0.0007 + 0.126 = 16.663[PO_4^{3-}] - 0.0007 + 0.0007$$

$$0.1267/16.663 = [PO_4^{3-}]$$

$$7.60 \times 10^{-3} \text{ M} = [PO_4^{3-}]$$

The other part of the question asks if this answer is reasonable. There are two reasons that we can say yes. First, the best-fit line to the data has a R^2 value of 0.9998, which is very close to 1.0, showing that our line equation is representative of the trend in the data. Second, the value we obtained from our calculation makes sense when we look back at the original data. The absorbance value we were investigating falls between 0.84 and 0.163 of our reference data, indicating that our concentration should fall somewhere between the concentrations of 5.00×10^{-3} M and 1.00×10^{-2} M, which it does.

This is therefore a good result. The process we just used to determine that this is an acceptable result illustrates an important principle: ***You should check to make sure that your calculated answers make sense whenever possible.***

Graphing: Using the Slope

Another type of graphical analysis you will be faced with in general chemistry is the extraction of data from the slope of a line. In particular, when we address the concept of kinetics, we can use this form of analysis to ascertain the individual orders of a rate equation. Reactions are categorized as zero-order, first-order, second-order, or mixed order (higher-order) reactions. These reaction orders are important because they tell us which reactant is most important in the overall rate by which a reaction progresses. We obviously don't expect you to understand all of these concepts at this point (there will be plenty of time for that later), but by determining the slope of the line resulting from a plot of the rate of a reaction versus the initial concentrations of each reactant we can determine the individual reaction orders for each one.

For example, let's say we want to determine the individual reaction orders for the reaction between nitrogen monoxide (NO) and oxygen (O_2) from the data presented below.

	$[NO]^o$ (mol/L)	$[O_2]^o$ (mol/L)	Instantaneous Rate (mol/L h)
Trial 1	0.020	0.010	0.028
Trial 2	0.020	0.040	0.114
Trial 3	0.020	0.020	0.057
Trial 4	0.040	0.020	0.227
Trial 5	0.010	0.020	0.014

What do we need to do? Well, first, we need a balanced chemical reaction:

$$2NO(g) + O_2(g) \Leftrightarrow 2NO_2(g)$$

Second, we need a rate equation:

$$\text{rate} = k[NO]^x[O_2]$$

Third, we need this equation rearranged such that it is in a linear form:

$$\log \text{rate} = x\log[NO] + \log k[O_2]$$

and

$$\log \text{rate} = x\log[O_2] + \log k[NO]$$

Please notice that the above equation is of the form $y = mx + b$, where x is the slope and reaction order in each case. You can look forward (if you would like) to the lab on kinetics to see a more complete breakdown of how these two equations were developed, but for our purposes here, we just need to know that they are linear equations.

At this point, two graphs must be generated, both with the form of log (rate) versus log (reactant). It is imperative that in the each graph one of the reactant's concentrations must be constant. Looking at the original data, this means that we will plot trials 3, 4, and 5 to generate our first graph ($[O_2]$ is constant), while trials 1, 2, and 3 will be used in the second graph ($[NO]$ is constant). Linear regression (e.g., adding a best-fit line) will allow us to determine the slope for each equation. Furthermore, x in both equations not only represents the slope of the line and but also corresponds to the individual reaction order for NO and O_2, respectively.

More data can still be gathered from our graphs. Observing the equations carefully, we notice the common factor of k, which, in short, is the rate constant. We can extract a value of k from each graph, average it, and report an overall rate equation.

$$\text{rate} = (7.4 \times 10^3 \text{ M}^{-2} \text{ s}^{-1}) [NO]^2[O_2]^1$$

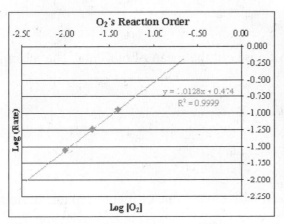 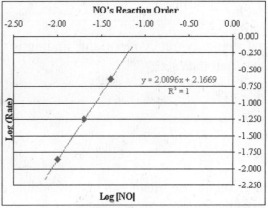

Example graphs of log [O₂] versus log (rate) and log [NO] versus log (rate) used to determine reaction orders

Statistics

Analyzing our data graphically is only half of the battle, as this type of analysis rarely gives information on how accurate or precise our data may be. For this aspect, we turn to statistical analysis. Statistics are a way for us to easily express the reliability of the data and analysis we present. The rest of the experiment will be directed toward introducing several statistical methods that must be mastered to express data in a manner that can easily be critiqued.

Overall, there are two areas of statistics of great importance to a chemist: the analysis of error in repeated measurements and the analysis of distributions and central tendencies. Basically, the first application deals with trying to ascertain the "true" value of a measurement, while the latter deals with observing patterns or trends in large collections of values or measurements.

Error Analysis

When you measure the same thing over and over again, you often end up with slightly different results—no matter how hard you try. This is solely due to the fact that every measurement has some error or uncertainty associated with it. Even advanced instrumentation such as radar guns have errors associated with them. To help understand how to handle this type of data, here are five measurements taken to determine the mass of carbohydrate in 50.0 g of a particular protein:

12.62 g, 11.91 g, 13.07 g, 12.73 g, and 12.59 g

If we were asked to report this scientifically, we would never just list the five values, we would give a "true" value and its associated error.

The Mean

The first step to correctly report our finding is to figure out the mean of our data. The mean is the average of our data set:

$$Mean(\bar{x}) = \frac{\sum_i x_i}{N}$$

For our example, the **mean** is found by adding each individual value or datum (x_i) and dividing by the size of the sample (N) as shown:

$$x = 12.58 \text{ g}$$

Standard Deviation

The next statistic we need to calculate is the standard deviation. Specifically, the standard deviation measures how closely data are clustered about the mean value and is technically defined as:

$$StDev(\sigma) = \sqrt{\frac{\Sigma(x - x_i)^2}{N-1}}$$

In general, the smaller the standard deviation, the closer the mean will be to the "true" value. In particular, the numerator in this equation calculates the residual for each piece of data. In other words, the numerator calculates how much an individual measurement differs from the mean. As for the squares and the square root, they take care of the fact that some of our data points are larger than the mean, while others are smaller than the mean.

NOTE: Some statistical approaches require the variance of the data. This is simply the square of standard deviation.

Again, using our example data:

$$StDev(\sigma) = \sqrt{\frac{(12.58 - 12.62)^2 + (12.58 - 11.91)^2 + (12.58 - 13.07)^2 + (12.58 - 12.73)^2 + (12.58 - 12.59)^2}{4}} = 0.4223$$

$$StDev(\sigma) = 0.42$$

Confidence and the Student's t

Finally, we can calculate the confidence (μ) with which we present our data. This statistical value incorporates several other values, including the mean, standard deviation in the mean, and the Student's t. This value is calculated by taking the mean and adding the corresponding confidence interval. The generalized form of the equation is shown below, where t is the value of the **Student's t** at a given number of degrees of freedom and confidence:

$$\mu = x \pm \frac{t \cdot \sigma}{\sqrt{N}}$$

For our example:

$$\mu = 12.58 \pm \frac{t \cdot 0.4223}{\sqrt{5}}$$

The value of t here is vital, as scientific data is generally expressed at 95% and 99% confidence. Using the table shown below and the fact that we have 4 degrees of freedom ($N - 1$), we see that our Student's t value will be either 2.78 or 4.60.

$$\mu_{95\%} = 12.58 \pm 0.053 \text{ g}$$

Or

$$\mu_{99\%} = 12.58 \pm 0.087 \text{ g}$$

Confidence table taken from *Analytical Chemistry: An Introduction,* 7th ed., Table 72, p. 152. Q Table taken from *Quantitative Chemical Analysis,* 5th ed., Table 4-5, p. 82.

Degrees of Freedom	90%	95%	99%
1	6.31	12.70	63.7
2	2.92	4.30	9.92
3	2.35	3.18	5.84
4	2.13	2.78	4.60
5	2.02	2.57	4.03
6	1.94	2.45	3.71
7	1.90	2.36	3.50
8	1.86	2.31	3.36
9	1.83	2.26	3.25
10	1.81	2.23	3.17
11	1.80	2.20	3.11
12	1.78	2.18	3.06
13	1.77	2.16	3.01
14	1.76	2.14	2.98
Infinite	1.64	1.96	2.58

Q Table (90% Confidence)	Number of Obs. (N)
0.76	4
0.64	5
0.56	6
0.51	7
0.47	8
0.44	9
0.41	10

What we have calculated is essentially error bars. We can report our data with 95% certainty that the amount of carbohydrate in this particular protein is within +/− 0.53 grams of 12.58 grams, while we know with 99% certainty that the amount of carbohydrate is within +/− 0.87 grams of 12.58 grams. In other words, we can conclude that this particular protein will contain ~12.00 to ~13.00 grams of protein at 95% confidence, and anywhere from ~11.40 to ~13.40 grams at 99% confidence.

Outliers

Utilizing our same example, what if we took one more experimental measurement and found the carbohydrate content to be 17.64 g. Immediately you notice that this value is well outside our confidence interval just reported, but being a good scientist, you include it in your calculations. Shockingly, you notice that its presence dramatically affects both your mean and standard deviation. Luckily, you are about to learn an approach with which you can statistically prove this data point to be an outlier, a piece of data that is "far away" from the rest of the data.

The Q-Test

The tool that will help you with this situation is the Q-test, a method dedicated entirely to determining if one particular data point can be rejected from the others. Shown below is the equation for $Q_{Calculated}$, which can be compared to a value from the Q_{Table}. If the value of $Q_{Calculated}$ is found to be greater than Q_{Table}, then the data can be discarded.

$$Q_{Calculated} = \frac{\text{gap}}{\text{range}}$$

For our example:

$$Q_{Calculated} = \frac{16.64 - 13.07}{16.64 - 11.91} = 0.755$$

In our example, $Q_{Calculated} > Q_{Table}$; we can therefore reject this data!

Report Contents and Questions

As the first experiment of the term, the exercises presented below are to be used to refresh your memory. None of this experiment is actually done in the laboratory. Therefore, since no data has been taken in the laboratory, none should be recorded in your notebook. Instead, this report should be prepared on notebook paper to be turned in at the next lab class meeting.

The report should consist of a title page, purpose, and the exercises below. Please remember all the rules for constructing good scientific graphs. Show all of your work for full credit.

Exercises

1) Consider the following experiment in which you wish to determine the molarity of the concentrated HCl in your new bottle of reagent. You carefully measure out 100.0 mL of the concentrated acid and dilute it to 1.000 L (in a 1 L volumetric flask). You then use a 5.00 mL volumetric pipette to remove 7 aliquots for titration with 0.1045 M sodium hydroxide solution. The following are the volumes of base needed for each sample.

   59.0 mL, 58.8 mL, 59.2 mL, 58.9 mL, 58.9 mL, 59.1 mL, 57.9 mL

   Calculate Q_{exp} for both the maximum and minimum values in this data.

   a. Q_{exp} for 59.2 mL =
   b. Q_{exp} for 57.9 mL =
   c. Can you reject 59.2 mL with 90% confidence?
   d. Can you reject 59.2 mL with 95% confidence?
   e. Can you reject 59.2 mL with 99% confidence?
   f. Can you reject 57.9 mL with 90% confidence?
   g. Can you reject 57.9 mL with 95% confidence?
   h. Can you reject 57.9 mL with 99% confidence?

   Average your measurements (throwing out the outlier if justified with 90% confidence) and calculate the molarity of the original concentrated HCl solution.

2) You are asked to calibrate a 10 mL volumetric pipette by weighing to the nearest 0.1 mg the mass of water delivered by the pipette. You weigh six samples of water delivered by the pipette and convert the mass of each to volume by dividing by the density of water at 25 °C (0.997048 g/mL). Following are your measurements:

   9.9690 mL, 9.9700 mL, 10.0360 mL, 9.9650 mL, 10.0150 mL, 10.0200 mL

   Calculate the following statistical measures for this data:

   Mean (x) = _____ Standard deviation (s) = _____

   Variance (s^2) = _____ Standard error of the mean (s_{mean}) = _____

   90% confidence interval = 9.9958 ± _____
   (The "true value" of the volume will lie within this interval 90% of the time.)

   95% confidence interval = 9.9958 ± _____
   (The "true value" of the volume will lie within this interval 95% of the time.)

99% confidence interval = 9.9958 ± _____
(The "true value" of the volume will lie within this interval 99% of the time.)

3) The table below shows some data in which the peak height of triplicate samples of a substance is listed as a function of the quantity of material injected. You would like to know first how good the precision of the method is by determining the standard deviations and confidence intervals of each set of triplicate samples. Then you would like to know whether the means of the triplicate samples show a linear relationship with sample size and the equation for this relationship.

Gas Chromatograph		Calibration	
Sample Mass (μg)	Peak Height (cm)		
	Trial 1	Trial 2	Trial 3
10	2.6	2.3	2.4
20	3.8	4.2	4.8
30	6.2	6.5	5.9
40	9.0	8.7	8.4
50	11.3	10.9	11.4
60	13.7	13.0	13.7

Q1. Using Excel® or another similar spreadsheet program, graph the average peak height versus sample mass. Apply a best-fit line to the resulting graph.

Q2. Again using the spreadsheet program, calculate *sample mean, variance, standard deviation, standard error of the mean*, and *confidence intervals* for the data above.

Q3. Answer the following questions:
   a. For the 60 μg sample, what is the sample standard deviation and 90% confidence interval of the three peak height measurements?

   b. Excel® gives the best straight line fit to the data as an equation $y = mx + b$. What are the slope and intercepts of this line? (Be sure to include units.)

4) The table below contains data that were collected to create a calibration curve in which the absorbance of a substance is plotted versus samples of known concentration. Graph this data.

Known Concentration (M)	Absorbance (AU)
0.10	0.115
0.20	0.260
0.30	0.485
0.40	0.790
0.50	1.175
0.60	1.640

Determine from your graph the concentrations of the following unknown solutions. (Give units.)

Solution A: 0.163 AU _____

Solution B: 0.658 AU _____

Solution C: 1.444 AU _____

5) The table below contains some initial reaction rate data for reaction of substances **A** and **B** at the indicated initial concentrations. The rate law for this reaction follows the form *rate* = $k[A]^a[B]^b$, where *a* and *b* are the **reaction orders** for **A** and **B**, respectively.

Experiment	Initial Conc. (M)	Initial Conc. (M)	Initial Rate (M/s)
	[A]	[B]	
Trial 1	2.0×10^{-2}	4.3×10^{-3}	2.2×10^{-3}
Trial 2	4.0×10^{-2}	4.3×10^{-3}	8.9×10^{-3}
Trial 3	6.0×10^{-2}	4.3×10^{-3}	2.0×10^{-2}
Trial 4	2.0×10^{-2}	8.6×10^{-3}	4.5×10^{-3}
Trial 5	2.0×10^{-2}	1.3×10^{-2}	6.7×10^{-3}

Q1. Plotting log rate versus log [A] at constant [B] will give *a* as the slope, and plotting log rate versus log [B] at constant [A] will give *b* as the slope. Use a spreadsheet to determine these values. Calculate the logarithm of the concentration and rate data to enter in the spreadsheet. Use trials 1, 2, and 3 for the log rate versus log [A] plot, and trials 1, 4, and 5 for the log rate versus log [B] plot. Add trend lines to the resulting graphs and determine the slopes of the lines.

   a. What is the reaction order for reactant A? _____

   b. What is the reaction order for reactant B? _____

Q2. The reaction rate constant (*k*) can be determined from the rearranged rate equation $k = [(rate)/([A]^a[B]^b)]$ using data from any of the trials. The units of *k* will depend on the **overall order** of the reaction, which in this case will be *a* + *b*. (The units of *k* must be such that the units of the rate become **M/s**.) The following table gives the units for different overall orders.

Overall Reaction Order (a + b)	Units of k
0	Ms^{-1}
1	s^{-1}
2	$M^{-1}s^{-1}$
3	$M^{-2}s^{-1}$
4	$M^{-3}s^{-1}$

What is the rate constant for the above reaction? (Include units.) _____

Experiment 2

A Submarine Adventure: Density Saves the Day

Introduction

Welcome to "A Submarine Adventure," the first "wet lab" in this manual (no pun intended). A "wet lab" in chemistry is simply a lab involving the use of chemicals. Although this lab will only expose you to water, you should realize that water is indeed a chemical and a very important one at that. The purpose of this lab, however, is not to focus on water, but rather to use water to illustrate a very basic physical concept, density. Density is calculated as the mass of an object divided by its volume ($d = m/V$). Density is an intensive property, meaning that it is a property that is the same no matter how much of a substance is present. Density is an important property to understand because it allows us to determine whether objects will float or sink when placed in a liquid or even a gas. Generally, substances float if their density is less than the density of the medium they are placed in. This is why small rocks sink to the bottom of a pond, while wooden objects, even large logs, float.

As scientists have discovered more about the concept of density, we (the scientific community) have learned to harness this knowledge for our own use. One of the most readily recognized uses of the concept of density is a modern invention: the submarine. Since steel is certainly much denser than water ($\sim$7.83 g/cm^3 versus $\sim$1.00 g/cm^3), one could ask the question, "How does a submarine that contains several tons of steel float in the ocean?" The answer is that a submarine takes advantage of the air space found within it. If the ship's total volume is much greater than its mass, then its density is less than that of the water around it, and it will remain afloat. To raise and lower the submarine in the ocean, the captain of the sub opens and closes compartments in the hull to allow water in and out and in doing so increases the submarine's mass; this causes the sub's density to increase and the submarine to submerge. To raise the submarine to the surface, the water is pumped out, reducing the submarine's density. The mechanical system that performs the influx and outflow of water into a submarine is a called a ballast system.

The purpose of this lab is to investigate the concept of density by determining the identity of various metal objects based on their experimentally determined densities and by building your own "mock" submarine. In doing so, you will learn some other very basic techniques that you will use again and again in the weeks to come. These techniques are the use of an analytical balance to weigh objects and the measurement of a liquid in a graduated cylinder.

Since a majority of the chemistry you will be completing in the lab is aqueous in nature, you will be measuring the volume of liquids in almost every lab. As with any process in science, there is a correct way to make the measurement. Please pay close attention to your instructor's demonstration of the proper way to fill and read your graduated cylinder. The other technique you will learn in this lab is the proper way to weigh an object. The analytical balance is a vital tool you will use to measure everything from solid reagents used to make solutions to products of a chemical reaction. An accurate weight is vital to both the production of accurate data and to the successful interpretation of an experiment. Your instructor will demonstrate the proper use of the analytical balance.

Background

Properties of Matter

Most likely you have already been introduced, either in the lecture for this class or in a prior science class, to the various properties used to characterize matter. Chemical properties of matter become apparent only during a chemical reaction. On the other hand, a physical property is an aspect of a sample that can be observed without changing its chemical composition. Examples include properties such as color and texture. Physical properties can be further described as extensive properties, those that depend upon how much of the substance is present (e.g., mass or volume), or intensive properties, those that are independent of the amount of substance present (e.g., temperature or density).

One of the most common physical properties of matter is mass, which is often referred to as weight. Although these two terms are often used interchangeably, they do not mean the same thing. Mass is used to describe the amount of matter that a substance contains, while weight is scientifically defined as the total gravitational force exerted on a substance. In fact, it is this notion that explains why astronauts are "weightless" in space, yet retain their mass. Another physical property that will be investigated in this experiment is volume. Like mass, the volume of a substance is an extensive property because it also depends upon how much of the substance is present.

Density

What is interesting about density is that it provides a direct relationship between two extensive properties, mass and volume. The density of a substance is usually reported in metric units that are a combination of the normal units used for mass, that is, grams (g) or kilograms (kg), and the normal units for volume, milliliters (mL), liters (L), or centimeters cubed (cm^3). For example, the density of solids and liquids is generally reported in g/cm^3 and g/mL, respectively, while for gases it is reported in g/L.

Density Calculations

Density is calculated as the mass per unit volume of a substance.

$$\text{density} = \frac{\text{mass}}{\text{volume}} \qquad \text{or} \qquad d = \frac{m}{V} \quad \text{(Equation 1)}$$

To calculate density, you must have two values: 1) the mass of the substance and 2) the volume of the substance. For example, calculate the density of a metal cylinder with a length of 2.75 cm and a diameter of 5.00 cm that has a mass of 422.8 g.

The volume of a cylinder is given by the equation $V_{cylinder} = \pi r^2 h$, where r is the radius of the cylinder and h is the height or length of the cylinder. Remembering that the radius of a circle is half the diameter, the calculation of the volume of the cylinder is:

$$V_{cylinder} = \pi (2.50 \text{ cm})^2 (2.75 \text{ cm}) = 54.0 \text{ cm}^3$$

To complete the calculation of density for the cylinder we simply substitute the value of the mass and volume into equation 1:

$$d_{cylinder} = \frac{m_{cylinder}}{V_{cylinder}} = \frac{422.8 \text{ g}}{55.7378227 \text{ cm}^3} = 7.59 \frac{\text{g}}{\text{cm}^3}$$

Seeing the resulting density of the metal cylinder, of which metal do you think it is composed? (Hint: read the introduction carefully.)

Exploring the Concept of Density through Experimentation

In this particular experiment you have two tasks: (1) to determine the identity of two unknown metal objects by matching their experimentally derived densities to a list of known metal densities and (2) to create a successful "submarine" by matching the density of a mock submarine (made of a balloon and metal pellets) to the density of the salt water in a mock ocean.

Identifying a Metal by Its Density

As you saw in the example above, once you have calculated the density of a metal object, that density can be used to identify the metal. As in the example, to experimentally determine the density of a metal object, you need to measure two physical properties of that object: its mass and its volume. Mass is simple to obtain. You will simply weigh the cylinder on an analytical balance, as described below. Obtaining the volume of your metal cylinders will require more work.

Determining Volume

To determine the volume of a liquid is easy; you simply place it in a graduated cylinder and record the resulting volume from the calibration on the side of the cylinder. To determine the volume of your metal objects, you have to use one of two methods: volume by displacement or volume by calculation. The first method, volume by displacement, uses the displacement of water in a graduated cylinder to determine the volume of the metal object:

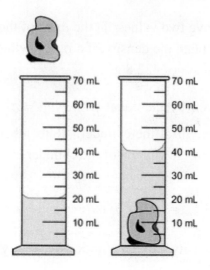

The volume of water that is displaced by the metal object (20 mL) is exactly equal to the volume of the metal object itself (20 mL).

Reading a Graduated Cylinder

Accurate measurement of this volume depends on the scientist's ability to correctly read the scale on the graduated cylinder. This reading is made more complicated by the formation of a meniscus caused by capillary action. Whenever liquids are held in a narrow container, the surface tension of the liquid causes a marked curvature of the upper surface. Specifically, this is referred to as a *meniscus*.

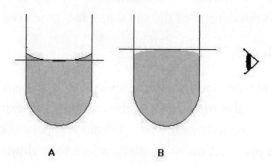

A: Read the bottom of a concave meniscus.
B: Read the top of a convex meniscus.

In the case of water, the meniscus is concave and to accurately record the volume, you have to observe at eye level where the center of the curve is located. At this point, it is appropriate to mention that to get the most accurate results, you also have to incorporate some estimation.

Graduated cylinders, thermometers, rulers, and other instruments are all read to the closest mark and then estimated to one decimal place further. This means dividing the space between marks into 10 smaller marks. For example, when reading a graduated cylinder with marks for each milliliter, you should record the volume to the nearest tenth of a milliliter. It is automatically assumed that the last number reported in your data is an estimate; this is also true for instruments that provide a digital output.

Volume by Calculation

The second method you can use to obtain the volume of your metal objects is called volume by calculation. Depending on the shape of your metal object, there are several equations shown below that may be used to calculate its volume:

For a cube $V = s^3$, where s is the length of the side in centimeters.
For a rectangle, $V =$ (length) $\times$ (width) $\times$ (height), each measured in centimeters.
For a cylinder, $V = \pi r^2 h$, where r is the radius of the cylinder and h is its length, both in centimeters.
For a sphere, $V = \frac{4}{3}\pi r^3$, where r is the radius of the sphere in centimeters.

You should note that each of these volume equations will result in the unit of cm^3. If your calculation does not result in this unit, you should check your calculations for errors.

The Vernier Caliper

For each of the equations above, you will need to measure the radius and/or the height (length) of your metal object. To measure these dimensions accurately, you will be using a device called a Vernier caliper.

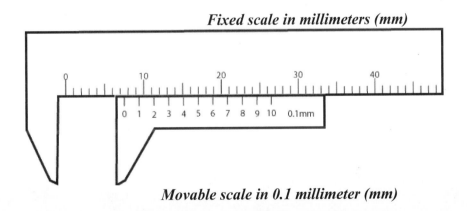

Fixed scale in millimeters (mm)

Movable scale in 0.1 millimeter (mm)

To use the instrument appropriately, please note the following guidelines:

1) Place the metal object in the caliper to measure your dimension.

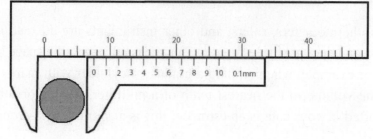

2) Close the caliper gently, but firmly, on the object so that the sides of the caliper are in firm contact with the object's sides.

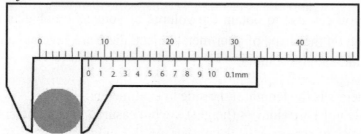

3) Find the 0 on the movable scale portion of the caliper. Looking just above and to the left of the 0, record, in millimeters (mm), the value of the non movable scale.

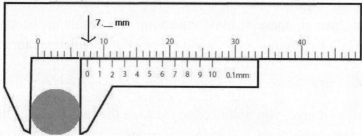

4) Now, find a line on the movable scale that completely coincides with a line on the non movable scale. Record that value to the nearest tenth of a millimeter (mm).

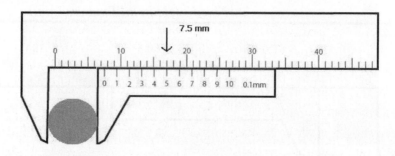

Measuring Mass

In addition to measuring the volume of your unknown metal object, to determine its density, you must also find its mass. There are many types of scales and balances, varying in range and accuracy, being used in today's laboratories and commercial operations. Different balances exist for different purposes; that is, you would not use a milligram balance to weigh out 2 tons of fertilizer.

All of the balances in the laboratory use the same general technique to determine the mass of an object. What follows is a brief introduction to each of the different types of balances commonly used:

1) Double-pan balance: Objects of an accurately known mass are added to one side of a two-pan scale to balance the object to be weighed until there is no difference between the two sides.

2) Mechanical balance: Similar to the double-pan balance; instead of placing weights on one side of the balance, internal weights inside the instrument are connected to dials on the front of the balance.

3) Electronic balance: A magnetic field is used to observe the deflection—which is compared to stored data for known standards—of the pan when an object is placed on it.

In a career as a scientist or engineer, all of these types of balances will be encountered. It would benefit you to become familiar with the ways in which each of them operates and to learn how to read the scales with the appropriate amount of precision and accuracy.

Now that you have all of the information you need on the measurement and calculation of density, you are ready to begin...

A Submarine Adventure

Picture this…you are at war with a group of aliens; we will call them Gatorians for lack of a better term. They look like big lizards with green skin and orange eyes. Yuck! They have taken over your world and have isolated you and 19 others on an island somewhere in the South Pacific. But you have hope of getting to safety. If you can rendezvous with the military in Hawaii, you and your friends will be protected. Your only problem is how to get there. The Gatorians patrol the ocean surface continuously, and Hawaii is hundreds of miles from your position. Then someone in your group suggests that you build a submarine. That way, you can avoid the Gatorians and still make it to Hawaii. You all decide that a submarine is the only way to go and then begin to design it. For the design to work, the submarine must float off the bottom of the ocean but not break the surface.

This means that the density of the sub as a whole has to be very similar to the density of the water that surrounds the island. Real submarines use ballast systems, of course, but you have limited time and supplies, so a design based on simple density is your only option (not to mention that none of you are physicists, and a workable ballast system is way out of your league). The group decides that they want you (the only scientist in the group) to design the sub and to make sure it will work. Before using what little resources you have to build the actual submarine, you devise the following experiment to test the idea first.

Using 20 metal pellets to represent the masses of the 20 people who have to travel inside the submarine and a balloon to represent the sub itself, you experimentally determine the density of the salt water and inflate the balloon to just the right volume so that the submarine's density allows it to hover between the bottom of a pail of water and the surface. Your model system is quite a bit smaller than the sub will have to be, but if you can get the model to work, then simply making the real sub 100 times bigger should also work as long as the density of the sub and the water remains the same.

Matching the Density of a "Submarine" to the Density of the Water around It

First Step:

You will need to experimentally determine the density of the "ocean water." This is done by gathering the mass and volume of a specific amount of the water. You should collect the following data in the lab: (1) graduated cylinder mass in grams; (2) graduated cylinder with ocean water mass in grams; and (3) volume in milliliters of ocean water.

The density of the ocean water is then calculated as:

$$density_{ocean\ water} = \frac{\text{graduated cylinder with water mass (g)} - \text{graduated cylinder mass (g)}}{\text{volume of ocean water (mL)}}$$

Your density is expressed in g/mL.

To match this density, you must figure out the mass and volume of the submarine you are making. The mass part is easy; you simply load the balloon with metal pellets and weigh them together. This is the mass of your submarine. The volume is a little bit more difficult, since that is what you will be varying to make the densities match.

The balloon is spherical in nature; thus, we can use the equation for the volume of a sphere, $\frac{4}{3}\pi r^3$, to calculate the density.

$$density_{ocean\ water} = density_{submarine} = \frac{\text{balloon and pellets mass (g)}}{\frac{4}{3}\pi r^3}$$

The *r* stands for radius, and that is what you will be solving for once you substitute all of your experimental numbers.

For example, if our calculated density of ocean water is 1.2 g/mL (or 1.2 g/cm³ this is because 1 g/mL = 1 g/cm³), and the mass of the balloons and pellets is 241.5 grams, then

$$1.2 \frac{g}{cm^3} = density_{submarine} = \frac{241.5 \, g}{\frac{4}{3} \pi r^3}$$

Solve for *r*, and you get **3.64 cm**. Since the diameter is twice the radius, this means you will need to blow your balloon up to a diameter, or width, of **7.28 cm**.

Procedure

SAFETY NOTES: In this and every other experiment, you must abide by the safety rules of your laboratory. Failure to comply with any of these requirements will result in your being asked to leave the laboratory.

Part I: Determining the Identity of a Metal by Its Density

Collect two unknown metal objects to investigate.

In your lab notebook, make careful observations about the markings, color and condition of the unknowns.

Use a standard weighing technique to weigh the metal unknowns on the analytical balance provided. Be sure to wipe the objects with Kimwipes® or an optical grade fiber cleaning wipe which leaves no residue to remove any fingerprints and so on. Record both masses in your lab notebook.

Use the appropriate method, volume by displacement or volume by measurement, to find the volume of your object. Be sure to note the method you chose and the reasons for your decision.

Using the density values you calculated in your prelaboratory assignment, determine the identities of your two unknown metal objects.

Part II: Building a Submarine

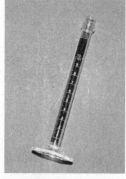

Weigh a 10 mL graduated cylinder using a standard weighing technique. Be sure to wipe it with Kimwipes to remove any fingerprints and so on. Record the mass in your lab notebook.

Use a disposable pipette to collect ~1–4 mL of ocean water from the reservoir provided. Transfer the ocean sample to the weighed graduated cylinder. Record the exact volume to 0.01 mL in your lab notebook.

Reweigh the graduated cylinder with the water in it using a standard weighing technique. Be sure to wipe it with Kimwipes to remove any fingerprints and so on. Record the total mass in your lab notebook.

Collect a balloon and 20 metal pellets from your instructor. Weigh the balloon and pellets using a standard weighing technique. Be sure to wipe them with Kimwipes to remove any fingerprints and so on. Record the total mass in your lab notebook.

Carefully place the pellets inside the balloon.

Using the calculated density of the ocean water and the equation given in the background section, calculate the diameter of the balloon needed to equal that density. Blow up your balloon to this diameter and seal it. (NOTE: Although the balloons should be round, they are often more elliptical in shape. Squeeze the balloon into a round shape as you measure it with your calipers for the best result.)

Test the viability of your submarine by placing it in the ocean. A good submarine will neither sink nor float; rather, it should hover just below the surface. Make notes in your lab notebook describing the results of this test.

Report Contents and Questions

The purpose should be several well-constructed sentences that describe the topics this experiment is designed to cover and the desired result. This should include both concepts and experimental techniques. The procedure section should reference the lab manual and note any changes that were made to the procedure while performing the experiment. Make sure to note which method you used when the procedure gives you a choice.

The data section should include all observations made during lab. The data table for determining the density of a metal should include the following information: **(a)** the measurements of the metal unknowns, **(b)** the volume of metal unknowns, **(c)** the mass of the metal unknowns, **(d)** the density of the metal unknowns, and **(e)** identity of metal unknowns. The second data table with the submarine data should include **a)** mass of graduated cylinder, **(b)** volume of ocean water, **(c)** mass of water, **(d)** mass of balloon with weights, **(e)** density of ocean water, **(f)** volume of balloon needed to match salt water density, **(g)** final volume of submarine design, and **(h)** final density of submarine design.

The calculations section should have example calculations of the following: **(a)** the volume of the unknown metals, **(b)** density of unknown metals, **(c)** percent error between known and unknown metal densities, **(d)** mass of water, **(e)** density of ocean water, **(f)** volume of balloon needed to match salt water density, **(g)** final volume of submarine, and **(h)** final density of your submarine.

The conclusion section should be in paragraph format. This section should report the results from the data section. For the density of the metal object section, be sure to report the density and identity of the unknown metals. Discuss any differences between the experimental and theoretical densities of the metals and the percent error. Also discuss any experimental errors that might have led to these differences. For the submarine data, report the density of the ocean water, the ideal volume of the submarine, and the final volume and density of the submarine. Discuss the trial and error portion of making the submarine and the differences in the ideal volume of the submarine and the final volume of the submarine. Finally, discuss any errors in the submarine design and the experiment.

Answer the following questions:

1) Give an example of the use of density (other than submarines) that you have observed in your own life.

2) What is third-person writing, and why is it used in the writing of scientific papers?

Experiment 2
Laboratory Preparation

Name: _____ Date: _____

Instructor: _____ Sec. #: _____

Show all work for full credit.

1) Read the background, procedure, and report sections of the lab experiment carefully and develop a hypothesis of what information you expect to gain from the completion of the lab experiment.

2) Create any and all tables you might need to collect data during the experiment and then transfer those tables into your lab manual for use during the lab.

Experiment 2
Prelaboratory Assignment

Name: _____ Date: _____

Instructor: _____ Sec. #: _____

Show all work for full credit.

1) A pycnometer with a mass of 56.96 g when empty has a mass of 108.22 g when filled with water (density = 1.000 g/mL) and a mass of 97.56 g when filled with liquid C.

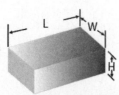
Empty pycnometer

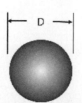
Pycnometer with water

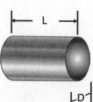
Pycnometer with liquid C

a. What is the volume of the pycnometer? _____

b. What is the density of liquid C? _____

2) Finding the density of solids requires a method of measuring the volume of the solid. If the solid has a regular geometric shape, the volume can be calculated from a measurement of the dimensions of the shape. Examples of three simple shapes are shown here, with the formulas for calculating their volumes.

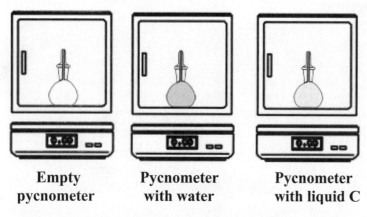

Rectangular Block

Sphere

Cylinder

$V = L \times W \times H$

$V = \frac{4}{3}\pi \left(\frac{D}{2}\right)^3$

$V = \pi \left(\frac{D}{2}\right)^2 \times L$

You measure the following dimensions of a rectangular metal block of metal A:
length = 10.89 cm; width = 6.49 cm; height = 1.57 cm. It has a mass of 193.07 g.

What is the density of metal A? _____

A sphere of metal B has a mass of 137.80 g and a diameter of 3.09 cm.

What is the density of metal B? _____

You measure the following dimensions of a cylinder of metal C:
length = 9.49 cm; diameter = 2.35 cm. It has a mass of 469.24 g.

What is the density of metal C? _____

Metal	Density * (g/cm^3)
Nickel	8.90
Lead	11.4
Aluminum	2.70
Brass	8.51
Copper	8.92
Iron	7.86
Magnesium	1.74

Based on the densities in the table, identify the unknown metals A, B, and C:

Metal A = _____

Metal B = _____

Metal C = _____

*McMurry & Fay 6th ed.

3) You determine that salt water in the tank has a density of 1.02 g/mL. The balloon weighs 2.0 g, and your weights have a mass of 30.0 g each. If you put six weights in your balloon, you must inflate the balloon to what diameter for it to have a density equal to the salt water and therefore float in the middle of the tank?

Experiment 3
Conservation of Matter

Introduction

There is a scientific law called the law of conservation of mass, discovered by Antoine Lavoisier in 1785. In its most compact form, it states:

Matter is neither created nor destroyed.

In 1842, Julius Robert Mayer published the law of conservation of energy. In its most compact form, it is now called the first law of thermodynamics:

Energy is neither created nor destroyed.

In the early twentieth century, Albert Einstein announced his discovery of the equation $E = mc^2$ and, as a consequence, the two laws above were merged into the law of conservation of mass-energy:

The total amount of mass and energy in the universe is constant.

What does this mean to us? Well, these laws allow us to balance chemical equations, calculate product amounts, and determine whether reactions will be spontaneous.

Our entire system of stoichiometry is based upon the veracity of these laws. The purpose of this lab experiment is to verify the first of these laws, the law of conservation of mass.

If you were to design an experiment to confirm this law, you would want to observe two things: (1) a reaction is taking place and (2) the total mass of all reactants is equal (within experimental error) to the total mass of all the products. To *observe* a reaction taking place, there must be a color change, the emission of a gas, or some other chemical change that can be visually monitored. There are reactions that we observe quite often, such as wood burning, that could be used but that are difficult to quantify because their products escape as soon as they are produced. When wood burns, it is converted into water and carbon dioxide, which escape as gases as soon as they are formed. More importantly, the other by-product of this reaction is extreme heat, which makes trapping the gases difficult and beyond the technology available in most introductory chemistry laboratories.

Since it produces both a color change and a gas, the reaction of copper(II) sulfate and zinc metal in aqueous HCl is useful for our purposes. The reaction can be monitored by observing the loss of blue color in the solution, the production of hydrogen gas, and the formation of solid copper. By quantifying the reactants and products of this reaction, we will be able to confirm that the total mass remains unchanged (within experimental error) while visually confirming that a reaction does in fact occur.

As always in lab, there are other purposes being served along with the learning of a new concept. We will also be revisiting the use of the analytical balance and producing our first chemical solution. Notice the phrase *within experimental error* that was used a couple of times above. This is because with any experiment, there is a certain amount of reactant and product lost when they are transferred from flask to flask or spilled, splashed, or dropped as part of the human error in the experiment. These "errors" must be taken into account when reporting the results of any experiment. Statistics are often used to indicate the relative importance of the error. For example, losing 100 g of product would seem tremendous unless it was compared to an expected product mass of 2.5×10^6 g. Then this error seems very small indeed. We will use this experiment to practice our knowledge and use of statistics to report the error in the mass of products created from the mass of reactants used.

Background

One of the most important aspects of chemistry is the balanced chemical equation. Recall from the purpose section that the law of conservation of mass states that matter can be neither created nor destroyed. Therefore, because matter is composed of atoms that are unchanged in a chemical reaction, the number and types of atoms in a chemical equation must be the same both before and after the reaction.

The Reaction

The experiment you are going to conduct involves the addition of zinc metal (Zn^0) to an acidic solution of copper(II) sulfate ($CuSO_4$), which is composed of copper(II) ions (Cu^{+2}) and sulfate ions (SO_4^{2-}). Thus, the reaction consists of the reactants, species on the left of the arrow: $Zn^0(s)$, $Cu^{2+}(aq)$, and $HCl(aq)$ in the form of $H^+(aq)$ and $Cl^-(aq)$, while the products of the reaction, species found to the right of the arrow, are $Zn^{2+}(aq)$, $Cu^0(s)$, and $H_2(g)$.

> You might notice that we don't mention the sulfate or chloride ions within the reactants or products. This is because in this reaction, they are "spectator" ions. They are called spectators because they don't participate in the reaction we are observing and remain at a constant concentration throughout the reaction. Just as in a math equation, any molecules or ions that remain the same on both sides of the reaction are said to cancel each other and can be removed from the net reaction equation. This doesn't mean those ions are gone, just that we can ignore them because they do not influence the reaction.

If we know the initial masses of both the solution and the metal pieces before the reaction, we can track how the masses have changed after the reaction, and we expect to observe that the total mass doesn't change. However, before we can proceed, we need to look at the chemical equations involved so that we can fully understand just what is taking place.

The reaction you will be running can be broken down into the two net ionic reactions shown below:

$$Zn^0(s) + Cu^{2+}(aq) \rightarrow Zn^{2+}(aq) + Cu^0(s)$$
$$Zn^0(s) + 2H^+(aq) \rightarrow Zn^{2+}(aq) + H_2(g)$$

Both of these reactions are referred to as oxidation-reduction reactions. We will return to this subject later, but for now it is simply important to note that the aqueous copper is being converted to solid metal copper. During the reaction, you will observe this process by the fading of the blue color characteristic of the Cu^{2+} ions and the accumulation of the pink copper metal on the zinc metal pellets.

Net Reaction

The two reactions above are then combined, like an addition problem, to form one equation:

$$Zn^0(s) + Cu^{2+}(aq) + 2H^+(aq) \rightarrow Zn^{2+}(aq) + Cu^0(s) + H_2(g)$$

In this equation, you can see that in addition to solid copper (Cu^0) being produced and solid zinc (Zn^0) being ionized to aqueous Zn^{2+}, hydrogen gas (H_2) is released.

Solution Preparation

The first task in this experiment requires you to make 25.0 mL of a 0.25 M solution of $CuSO_4$ in 3.0 M HCl. Because this is one of the first solutions you will be preparing this term, what will follow is a tutorial-like discussion of the proper method for preparing solutions.

As an example, let's pretend that we have to make 50.0 mL of a 0.75 M solution of potassium permanganate ($KMnO_4$) in 3.0 M HCl. The first step in this process is to obtain about 50 mL of the 3.0 M HCl in your 100 mL graduated cylinder. Next, we have to calculate how many grams of $KMnO_4$ we will need to make a 0.75 M solution. To do this, we have to find the number of moles needed for our solution and then multiply by the formula mass of $KMnO_4$.

The molar mass for a compound is calculated by adding together all of the molar masses found in the formula. For $KMnO_4$ this is the mass of 1 potassium atom + the mass of 1 manganese atom + the mass of 4 oxygen atoms:

$$\text{molar mass } KMnO_4 = 39.0983 \tfrac{g}{mol} + 54.938049 \tfrac{g}{mol} + 4(15.9994 \tfrac{g}{mol}) = 158.03 \tfrac{g}{mol}$$

Molarity is defined as the number of moles of solute divided by the liters of solution. Note that we said liters of solution, not just liters of solvent. This means that the total solution volume (including solute volume) is in the denominator ($M = mol/L_{soln}$).

For our solution, the solute is the potassium permanganate. Since we know the molarity of the solution is supposed to be 0.75 M and the volume is supposed to be 50.0 mL, we can calculate the number of moles of $KMnO_4$ needed:

$$moles_{KMnO_4} = \frac{0.75\, mol\, KMnO_4}{L_{KMnO_4\, solution}} \times 0.050\, L = 0.0375\, mol\, KMnO_4$$

$$\mathbf{moles_{KMnO_4} = 0.038\, moles}$$

The hardest part is now done! With the number of moles of $KMnO_4$ known, all we have to do is multiply by the molar mass of potassium permanganate to get the number of grams.

$$0.0375\, mol\, KMnO_4 \times \frac{158.03\, g\, KMnO_4}{1\, mol\, KMnO_4} = 5.93\, g\, KMnO_4$$

$$\mathbf{grams_{KMnO_4} = 5.9\, g}$$

We now need to prepare the flask in which we will make the solution. Solutions are normally made in specially calibrated volumetric flasks. For our solution, we will use a 50.0 mL volumetric flask, filling it about two thirds full with the 3.0 M HCl we are using as the solvent for our solution. Note that you should never completely fill the flask before adding the solute. This is because addition of the solute will affect the overall volume of the solution, and you don't want to exceed the 50.0 mL total volume. Next, we weigh out the 5.93 g of $KMnO_4$ using the proper weighing technique and add this amount to the 3.0 M HCl in the flask. After stirring, we can add HCl so that the final total solution volume is 50.0 mL. We will use this same process to produce the solution needed to perform this experiment.

Volume of H_2 Evolved

As stated above, the reaction we will be observing produces hydrogen gas. To determine that the mass of reactants and products has been conserved, we need a method by which we can collect and determine the mass of this hydrogen gas. To accomplish this task successfully, you will use a **side-armed Erlenmeyer flask**:

and a balloon. By sealing the end of the balloon to side-arm, all of the gas released from the reaction will fill up the balloon. We can then determine the volume of the balloon by measuring its radius (r), and height (h) if necessary, and substituting them into one of the following equations:

$$\text{Vol}_{sphere} = \frac{4}{3}\pi r^3 \qquad \text{or} \qquad \text{Vol}_{cyl} = \pi r^2 h$$

Assuming the temperature of the gas is approximately room temperature, 298 K (or 25 °C) and that the atmospheric pressure is 1.00 atmosphere, you can calculate the mass of hydrogen gas by substituting the volume of hydrogen gas in L into the following equation:

$$\text{mass of } H_2 \text{ gas (in g) evolved} = 8.244 \times 10^{-2} \times (\text{volume in L})$$

(**Reminder:** $1 \text{ cm}^3 = 1 \text{ mL} = 0.001 \text{ L.}$)

Procedure

SAFETY NOTES: The hydrochloric acid (HCl) used in this lab is very concentrated and can cause severe burns. If the acid comes in contact with your skin, rinse the area immediately with lots of water.

Part I: Making the Copper(II) Sulfate Solution

You will need to create 25 mL of a 0.25 M solution of CuSO$_4$ in 3 M HCl. Using your 100 mL graduated cylinder, collect ~25 mL 3.0 M HCl from the designated area and place it in a beaker. Be careful!

$0.025(3) = 0.075$ mol

$ML = mol$ $M = \frac{mol}{L}$

$0.025(3) = mol$

Using the weigh boats provided, weigh out the correct amount of $CuSO_4$ you calculated in your prelaboratory assignment. Be sure to use proper weighing technique.

Fill the volumetric flask about one third full with 3.0 M HCl. Add a small amount of HCl to the weigh boat containing your $CuSO_4$. Slowly pour the HCl/$CuSO_4$ slurry into the volumetric flask. Add HCl to the $CuSO_4$ in the volumetric flask until the volume is 25 mL exactly, as indicated by the white line on the flask. Allow the solution to return to room temperature. Check to see that the volume is still correct. Add a little more HCl if necessary.

Part II: Precipitating Cu^{2+} Ions as Copper Metal

Collect and weigh all of the following items and record the masses in your notebook to the nearest 0.0001 g.
1) A 100 mL side-arm flask
2) Four or five pieces of solid zinc metal (weigh together on a clean Kimwipe)

Place 25 mL of the copper sulfate solution in the side-arm flask and reweigh. Record the mass in your lab notebook to the nearest 0.0001 g.

Collect a balloon from the front counter and fit it onto the side-arm of the preweighed flask. Using a rubber-band or parafilm, make sure the seal between the balloon and the sidearm is airtight.

Collect a stopper from the front counter. You can add parafilm to the flask before adding the stopper to guarantee a good seal.

You are now ready to start the reaction. Quickly place the preweighed zinc metal pieces into the solution and cap with the stopper. You should see bubbles form right away. This is the hydrogen gas being produced.

Make observations about how the solution's appearance changes as the reaction proceeds, noting if there are any color changes. Make sure the balloon stays on the side-arm of the flask, as increased gas pressure might push it off.

When the reaction is complete (no more bubbles and the solution should be clear and colorless), you will need to carefully measure the balloon. Record the diameter in your notebook.

Make observations about the solution and the metal pellets in the flask. Specifically, detail how these observations support the claim that a reaction took place.

Finally, remove the balloon, stopper, and parafilm from the flask. Reweigh the flask and its contents and record the weight in your notebook to the nearest 0.0001 g or the smallest number your scale can read.

Dispose of the solution and pellets in the appropriate waste container(s).

Report Contents and Questions

The purpose should be several well-constructed sentences describing what your experiment was designed to accomplish and the criteria used to determine success. These sentences should include both concepts and techniques. The procedure section should reference the lab manual and include any changes made to the procedure during the lab.

The data section should include a table with the following information in a well-organized format. Initial data should include **(a)** mass of the 125 mL side-arm flask, **(b)** mass of four or five pieces of solid zinc metal, **(c)** mass of 125 mL side-arm flask with $CuSO_4$ acidic solution, **(d)** mass of $CuSO_4$ acidic solution alone, and **(e)** total mass before the reaction. Final data should include **(a)** mass of 125 mL side-arm flask with $CuSO_4$ acidic solution, **(b)** mass of solution alone, **(c)** volume of balloon, **(d)** calculated mass of H_2 gas alone, and **(e)** total mass after the reaction. Finally, calculate the mass percent recovered after the reaction.

$$\text{mass\%}_{recovered} = \frac{\text{total mass after reaction}}{\text{total mass before reaction}} \times 100\%$$

Most of the information for the tables above can be copied into the report right out of your lab notebook. The data section of your report should also include all observations made during the experiment.

For the calculation section, show one sample calculation for each calculation done to complete the data tables above. Make sure that this section is typed or written in ink and that each calculation is labeled as to what is being calculated and why.

The conclusion for this experiment should be in paragraph format. State the total mass before the reaction and the total mass after the reaction, then discuss the mass percent recovered. Be sure to explain whether your experiment confirms the law of conservation of mass and provide evidence for your conclusion. Also discuss the mass of the H_2 gas the reaction produced. Finally, include a discussion of possible errors in the experiment and how they might have affected the results.

Answer the following questions:

1) Why is the acid in the reaction necessary? Why can the solution of copper(II) sulfate simply be made with water?

2) Explain how the law of conservation of mass is the basis for all of stoichiometry.

Experiment 3
Laboratory Preparation

Name: _____ Date: _____

Instructor: _____ Sec. #: _____

Show all work for full credit.

1) Read the background, procedure, and report sections of the lab experiment carefully and develop a hypothesis of what information you expect to gain from the completion of the lab experiment.

2) Create any and all tables you might need to collect data during the experiment and then transfer those tables into your lab manual for use during the lab.

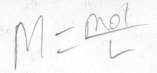

$$M = \frac{mol}{L}$$

3) In your laboratory, you will be required to make a solution of 0.25 M $CuSO_4 \cdot 5H_2O$ in a 25.00 mL volumetric flask. This exercise will walk you through the steps involved. After completion, write down the values for your answers in your notebook and take them to the laboratory to use in preparing your solution.

159 g
2509

Q1. What is the formula weight of $CuSO_4 \cdot 5H_2O$?

.025 L

Q2. How many moles of $CuSO_4 \cdot 5H_2O$ are required to make 25.00 mL of a 0.25 M solution?   .00625 mol

Q3. What mass of $CuSO_4 \cdot 5H_2O$ will be needed to make this solution?
1.5625   0.99375 g

Q4. You will add the salt to a weigh boat, which you determine has a mass of 2.1397 g. What will be the reading on the balance when you have put sufficient $CuSO_4 \cdot 5H_2O$ in the boat?

Experiment 3
Prelaboratory Assignment

Name: _____ **Date:** _____

Instructor: _____ **Sec. #:** _____

Show all work for full credit.

1) Calculate the number of moles or grams in each of the following quantities of salt.

   8.745 g of $Cu(NO_3)_2 \cdot 6H_2O$ =   _____

   3.545 g of zinc nitrate trihydrate =   _____

   2.516×10^{-2} mol of $Na_2SO_4 \cdot 10H_2O$ =   _____

   1.198×10^{-2} mol of zinc nitrate hexahydrate =   _____

2a) You would like to make up 745 mL of a 0.23 M solution of $ZnSO_4 \cdot 7H_2O$. How many moles of the salt do you need? How many grams of the salt do you need?

2b) You have a 0.43 M solution of $Na_2SO_4 \cdot 10H_2O$. What volume of the solution (in milliliters) must you measure to have 0.20 moles of the salt?

2c) You have a 0.57 M solution of $Cu(NO_3)_2 \cdot 6H_2O$.
What volume of the solution (in milliliters) must you measure to have 98.6 g of the salt?

3) Zinc metal reacts with acid solution to produce hydrogen gas and zinc ion as follows:

$$Zn(s) + 2H^+(aq) \rightarrow Zn^{2+}(aq) + H_2(g)$$

You add 6.825 g of Zn metal to 29.0 mL of 1.70 M HCl, and set up the flask to catch the hydrogen in a balloon, as in the following picture.

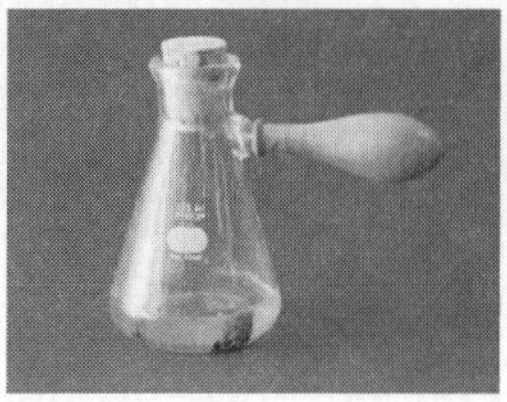

How many moles of Zn have you added to the flask?

How many moles of H^+ are in the solution?

How many moles of Zn will have reacted with the H^+ when the reaction is complete?

How many moles of Zn will remain unreacted?

What mass of Zn will remain unreacted?

How many moles of H_2 gas will be produced?

What mass of H_2 gas will be produced?

Assuming that one mole of H_2 gas would occupy a volume of 22 L at the temperature and pressure in the balloon, what would be the volume of the balloon?

Experiment 4
Plop, Plop, Fizz, Fizz: The Mass Percent of NaHCO$_3$

Introduction

As we continue through lab this semester, we will revisit the concepts of stoichiometry and mole–mass relationships. One of the relationships that strengthens our understanding of both the math involved in and the relevance of stoichiometry is mass percent. The mass percent of a compound or element within another compound or mixture is often vital to the viability of a reaction. For example, in a hydrate, the mass of the salt portion of the molecule is often a low percentage of the overall molecular mass. Magnesium chloride hexahydrate (MgCl$_2$•6H$_2$O) is such a compound. The overall molecular mass is 203.3 g/mol, but the salt is only ~47% of that mass. If I were manufacturing this compound, I would have to take that into account because a majority of the product mass would not contain the salt compound I need.

Another place where mass percent is important is in medications. Very often, the active ingredient in a medication is a very small portion of the overall mass of the tablet or capsule. The rest of the tablet is generally binders, flavors, and sometimes other medications designed to counteract side effects. The Alka-Seltzer® tablets that we will investigate in this lab contain not only sodium bicarbonate, the acid neutralizing ingredient, but also two other ingredients, aspirin and citric acid. Our purpose in this lab is to determine the mass percent of the sodium bicarbonate in an Alka-Seltzer tablet and compare our results to the published mass percent given by the manufacturer.

Because the sodium bicarbonate cannot be measured directly as it reacts with the other ingredients in the tablet, we will measure the production of carbon dioxide and use stoichiometry to calculate the initial mass of the sodium bicarbonate. This will also strengthen your understanding of mass to mole relationships. Using a balanced chemical reaction, any chemist should be able to tell you how much product can be made from a given amount of reactant or how much reactant was used to produce a particular amount of product.

Background

Everyone is probably familiar with the old Alka-Seltzer commercial starting with "Plop, plop, fizz, fizz, oh what a relief it is…" However, you probably have never thought about how Alka-Seltzer provides its relief. According to its packaging, an Alka-Seltzer tablet contains 1916 mg of sodium bicarbonate (NaHCO$_3$), 325 mg of acetylsalicylic acid (HC$_9$H$_7$O$_4$), and 1000 mg of citric acid

($H_3C_6H_5O_7$). When the tablet is dissolved in water, the sodium bicarbonate reacts with acetylsalicylic and citric acid according to the following chemical reaction:

$$2NaHCO_3(s) + HC_9H_7O_4(s) + H_3C_6H_5O_7(s) \rightarrow 2H_2CO_3(aq) + NaC_9H_7O_4(aq) + NaH_2C_6H_5O_7(aq)$$

Though the above chemical equation appears absolutely terrifying at first glance, it is rather simplistic once we look in greater detail at what is actually reacting. Each reactant behaves as a weak electrolyte. This means that when dissolved in water, each species breaks down into its component ions. For example, the sodium bicarbonate ($NaHCO_3$) breaks down into a sodium ion (Na^+) and bicarbonate (HCO_3^-), while the acetylsalicylic acid breaks down to H^+ and $C_9H_7O_4^-$ and the citric acid decomposes to H^+ and $H_2C_6H_5O_7^-$. The species holding the key to the relief of an Alka-Seltzer tablet is the bicarbonate ion. When dissolved in water, the sodium bicarbonate undergoes an acid–base reaction with the two acids, which for the sake of simplicity will now just be referred to as H^+. Thus, our frightening equation above is dramatically simplified to:

$$HCO_3^-(aq) + H^+ \rightarrow H_2CO_3(aq)$$

The product in this equation, carbonic acid (H_2CO_3), is extremely unstable and quickly breaks down into water (H_2O) and carbon dioxide (CO_2) gas. Thus, our chemical equation now takes the form:

$$HCO_3^-(aq) + H^+ \rightarrow H_2O(l) + CO_2(g)$$

This release of CO_2 gas is what produces the bubbles seen when we add the tablet to water. Since the CO_2 molecule was originally part of the mass of the Alka-Seltzer tablet, its release into the atmosphere should result in a net loss of mass after the reaction is complete. Thus, if we know the mass of the water and tablet before combining them, we can subtract the mass of the water and the tablet after the reaction is complete to get the mass of CO_2 lost. Since there is a 1:1 ratio of CO_2 to HCO_3^- in the reaction, we can then calculate the amount of $NaHCO_3$ reacted and therefore finally determine the mass percent of this species in our Alka-Seltzer tablet.

Ideally, we hope that all the sodium bicarbonate would react; however, this is not the case because some is always left over. This is because the mole amounts of the acids in each tablet are deliberately less than the mole amount required to react with all of the $NaHCO_3$. The purpose of this excess is to make sure that there is always enough sodium bicarbonate available to neutralize the stomach acids responsible for heartburn and stomachaches. In this experiment, acetic acid ($HC_2H_3O_2$) will be used to simulate our stomach juices. Overall, the experimental reaction is now:

$$HC_2H_3O_2(aq) + NaHCO_3(s) \rightarrow NaC_2H_3O_2(aq) + H_2CO_3(aq) \rightarrow H_2O(l) + CO_2(g) + NaC_2H_3O_2(aq)$$

To determine the amount of excess $NaHCO_3$, Alka-Seltzer tablets will be added to eight different solutions with increasing amounts of acid. The amount of CO_2 produced by the reaction of $NaHCO_3$ with $HC_9H_7O_4$ and $H_3C_6H_5O_7$ can then be determined as previously described. With the mass of carbon dioxide known, stoichiometry can then be utilized to work backward to determine the mass of $NaHCO_3$ reacted in each trial. The mass percent is calculated for each trial with a plot of mass percent versus volume of vinegar being generated to experimentally determine the total amount of sodium bicarbonate in an Alka-Seltzer tablet.

$$mass\% = \frac{mass_{NaHCO_3}}{mass_{Tablet}}$$

Procedure

SAFETY NOTES: While the acetic acid being used is not a strong acid, it can cause irritation if splashed in the eyes or left in prolonged contact with your skin. The Alka-Seltzer can react violently enough to "spray" some acid out of the beakers, so wear goggles at all times.

GENERAL INSTRUCTIONS: Students should work in teams to complete this experiment, with each student completing two of the eight runs required.

Weigh a clean, dry 100 mL beaker and small watch glass to the closest milligram (0.001 g). Record the weight in your lab notebook.

Using a graduated cylinder, place 35 mL of de-ionized distilled (DI) water in the 100 mL beaker you weighed.

Re-weigh the beaker with the water. Make sure the outside of the beaker is dry and use Kimwipes to remove any fingerprints, dust, and so on. Record the weight in your lab notebook.

Collect an Alka-Seltzer tablet. Break it in half and weigh each half carefully to the closest milligram.

Add one of the half-tablets to the beaker with the water in it. Place the preweighed watch glass over the top of the beaker (place the watch class over the beaker inverted as shown in the image to the left) to deflect any splatter. Swirl the contents to ensure the tablet is dissolved completely into solution.

After the reaction is complete (stopped bubbling), record the mass of the beaker, its solution, and the watch glass. Record the mass in your lab notebook.

Empty the contents of the beaker down the sink and rinse the beaker with DI water. Dry the beaker and watch glass thoroughly using paper towels and Kimwipes.

Using a graduated cylinder, place 5 mL of vinegar in the 100 mL beaker. Then add water until the volume in the cylinder is 35 mL.

Place the solution into the previously weighed beaker.

Reweigh the beaker with the solution. Make sure the outside of the beaker is dry and use Kimwipes to remove any fingerprints, dust, and so on. Record the weight in your lab notebook.

Add the second half-tablet to the beaker with the water in it and cover with the watch glass (make sure the watch glass is dome side up) as before. Swirl the contents to ensure the tablet dissolves completely into the solution.

After the reaction is complete, reweigh the beaker, its solution, and the watch glass. Record the weight in your lab notebook.

Repeat the experiment for 10, 15, 20, 25, 30, and 35 mL of vinegar.

Report Contents and Questions

The purpose should be two or three sentences stating why this lab was done. This should include but not be limited to key concepts and techniques. Be sure to include multiple topics within the purpose, since each lab covers more than one topic. The procedure section for this experiment should cite the lab manual and include a note with any changes made to the experiment during lab. Be sure to include the names of your team members and which trials with acetic acid were performed by each person.

A data table similar to the example below should be completed and presented in the data section.

Title

Volume			Mass before Reaction (g)				Mass after Reaction (g)	CO_2		$NaHCO_3$		
RUN	VINEGAR	WATER	½ TABLET	BEAKER	Beaker and Solution	Watch Glass	Total Mass	Beaker and Solution and Watch Glass	GRAMS	MOLES	MOLES	GRAMS	MASS %
1	0	35											
2	5	30											
3	10	25											
4	15	20											
5	20	15											
6	25	10											
7	30	5											
8	35	0											

The mass before reaction and mass after reaction sections should have all been collected in your lab notebook and copied into your lab report. Be sure that you have the correct number of significant figures, based on the balance you used.

You can use your stoichiometry skills to calculate the CO_2 and $NaHCO_3$ sections. First, determine the correct balanced chemical equation from the background section. Write this equation in your data section above your data table. To determine the grams of CO_2, subtract the total mass after the reaction from the total mass before the reaction. Convert the grams of CO_2 into moles of CO_2 using the molar mass of CO_2. Then, using the mole ratio from the balanced chemical equation, find the moles of $NaHCO_3$. Report the grams of $NaHCO_3$ by converting the moles into grams using the molar mass of $NaHCO_3$. Finally, determine the mass percent of $NaHCO_3$.

For the calculation section, show one sample calculation for each calculation done to complete the data table. Make sure that this section is typed or written in ink. This section should also include the calculation for the percent error in the mass percent for each run. This can be determined by using the information provided in the background section regarding the known amount of $NaHCO_3$ in a tablet. Be sure that all of your units are correct before substituting them into the following equation:

$$\% \text{ error} = \frac{\text{actual Value} - \text{experimental value}}{\text{actual value}} \times 100\%$$

Also include an Excel graph of the mass percent of $NaHCO_3$ versus milliliters of acetic acid. The mass percent of $NaHCO_3$ goes on the y-axis, and the volume of acid on the x-axis. Remember all graphing guidelines apply here. Use where the graph begins to level out to determine the total amount of $NaHCO_3$ in a tablet.

The conclusion section should be several paragraphs in length and cover the following topics: a discussion of the trends in mass of CO_2, mass of $NaHCO_3$, and mass percent found in your data table. Be sure to include values in your discussion. This section should also state the minimum volume of acid required to react all of the $NaHCO_3$ present in an Alka-Seltzer tablet. The minimum volume you pick should be supported and explained using your data. Be sure to state and explain the total amount of $NaHCO_3$ in an Alka-Seltzer tablet. A discussion of the percent error in each run should be included, as well as any errors in the experiment that may have led to the discrepancies in values.

Answer the following questions:

1) Explain why the mass percent in a half-tablet is the same as the mass percent in a whole tablet of Alka-Seltzer.

2) Explain how the graph of mass percent of $NaHCO_3$ versus milliliters of acetic acid helps you determine the amount of $NaHCO_3$ in an Alka-Seltzer tablet.

Experiment 4
Laboratory Preparation

Name: _____ Date: _____

Instructor: _____ Sec. #: _____

Show all work for full credit.

1) Read the background, procedure, and report sections of the lab experiment carefully and develop a hypothesis of what information you expect to gain from the completion of the lab experiment.

2) Create any and all tables you might need to collect data during the experiment and then transfer those tables into your lab manual for use during the lab.

Experiment 4

Prelaboratory Assignment

Name: _____ Date: _____

Instructor: _____ Sec. #: _____

Show all work for full credit.

1) Acidification of sodium bicarbonate ($NaHCO_3$) produces carbon dioxide in a two-step process. The first step produces carbonic acid (H_2CO_3). Write and balance the molecular equation for the first step of the reaction of sodium bicarbonate with HF.

The second step is the decomposition of carbonic acid to carbon dioxide. Write and balance the molecular equation for the second step.

Now, write the overall net ionic equation for the acidification of $NaHCO_3$ to produce CO_2.

What is the mole ratio of CO_2 produced to $NaHCO_3$ decomposed?

What mass of CO_2 will be produced by the complete decomposition of 1346 mg of $NaHCO_3$?

2) Two enterprising chemistry students decide to develop an alternative to Alka-Seltzer tablets, replacing the aspirin (acetylsalicylic acid) with vitamin C (ascorbic acid). The following table shows one of the formulations they tried for a 5 g tablet.

Ingredient	Formula	Quantity (mg)
Sodium bicarbonate	$NaHCO_3$	1991
Citric acid	$H_3C_6H_5O_7$	886
Vitamin C	$HC_6H_7O_6$	370
Inert binder		1753
Total mass:		**5000**

When the tablet dissolves in water, it "fizzes" as the bicarbonate reacts with the citric acid and the vitamin C. What mass of the $NaHCO_3$ reacts with the vitamin C?

Citric acid is triprotic and reacts with base to form the citrate ion ($C_6H_5O_7^{3-}$).
Write the overall net ionic reaction for the acidification of $NaHCO_3$ by citric acid, giving CO_2 as the product. Citric acid should be written as $H_3C_6H_5O_7$ in the net ionic equation, as weak acids are written in the molecular form even though they slightly ionize.

What mass of the $NaHCO_3$ reacts with the citric acid?

What is the mass of CO_2 released in this reaction?

The remaining $NaHCO_3$ is available to neutralize stomach acid. What mass of stomach HCl could be neutralized by one tablet?

Experiment 5
Proof of Alcohol

Introduction

What is the meaning of the word *proof* when referring to alcoholic beverages? Although most students know that the higher the proof of a liquor the greater its "punch," they are generally unaware of the term's true meaning. This experiment was designed to answer that question. As well as to teach two important laboratory techniques: the use of a separatory funnel and the use and care of Pyrex® glassware for handling liquids. Other topics addressed are concepts involved with separating and extracting materials using immiscible liquids; working with concentration units in volume percent, mass percent, and proof; and the patent process.

Background

Use of a Separatory Funnel

A separatory funnel is a standard piece of equipment used in synthetic chemistry labs. Generally made of glass, the funnel has a conical body, a glass or Teflon® stopcock to control the release of solution from the bottom of the funnel, and a glass or Teflon stopper at the top.

The stopper must be removed to release the liquid from the funnel. If the stopper is not removed, the vacuum that forms above the liquid will prevent the solution from draining properly. The vacuum will remove the air (from the stem) and the bubbles that form will cause the phases to mix again, defeating the purpose of the separation.

Glass stoppers and stopcocks should be greased to produce a good seal. Teflon joints should not be greased.

Performing an Extraction

1) Place the separatory funnel in a support (generally a ring stand). Remove the stopper and make sure that the stopcock is closed.

2) Add the solution to be extracted. The funnel should not be filled more than half full. Place the stopper on top. There should still be some room.

3) Take the separatory funnel out of the support and hold it tightly at the stopper and the stopcock. Invert it slowly and vent (open the stopcock) away from yourself and others. You will hear a kind of whoosh when the pressure is released.

4) Close the stopcock and invert the funnel gently. Vent it again. Repeat this step until no more gas escapes.

5) Place the separatory funnel back into its support. Allow the layers to separate. Then remove the stopper and drain the bottom layer into a clean container. At this point, you need to know which layer contains your desired product. The layers separate due to immiscibility and density. The less dense liquid will reside in the top layer. As organic liquids are generally less dense than water, aqueous solutions will generally form the bottom layer in a separation.

History

In the 1980s, a patent was developed that separated ethanol from water by separatory funnel, rather than by distillation. The process involves the use of several chemical reactions and techniques that cannot be recreated exactly in the chemistry teaching lab, but the basic concepts of the patent can be demonstrated by separating a mixture of water and ethanol. At first, the ethanol will be miscible with the water, forming a homogeneous mixture, but upon addition of an organic solvent that binds ethanol, it will separate as the upper layer and can be collected. When optimized, a process like this could have profitable use in beer, wine, or even pure ethanol industries. Until that optimization is completed, however, the inability to complete the patent process in its entirety leads to a less than 100% efficiency of separation.

Separation of immiscible liquids is a concept that requires an understanding of homogeneous and heterogeneous mixtures, so these concepts are incorporated into this experiment. Since ethanol is miscible in water before addition of diethyl ether and sodium acetate, it can be seen during the experiment that a molecular interaction is occurring that causes the ethanol to separate.

Volume Percent, Mass Percent, and Proof (Ethanol)

Concentration is important in all of chemical and physical understanding. Knowledge of volume percent and mass percent will always be necessary when gathering and analyzing quantitative data, especially for liquids. The distinction between volume percent (which is dependent on temperature and pressure) and mass percent (which is independent of temperature and pressure) is made and exemplified in this experiment. The meaning of *proof* (when referring to ethanol in water) can be demonstrated and in normal daily life has usefulness if one is consuming alcoholic beverages.

The concept of concentration is very wide and diverse, though it remains one of the most useful and common tools in the chemist's box of skills. Consider a pure material and an impurity; when only the pure material exists, its concentration is maximized. With the addition of an impurity, a change in concentration occurs as the system (pure material + impurity) gets bigger. With continued addition of the impurity, the concentrations will continue to change respectively for each material. Concentration units such as molarity, volume percent, and mass percent, where the quantity of the species for consideration is the numerator and the quantity of total system is the denominator for the fractional ratio are used to describe the relationship between the pure substance (solvent) and the impurity (solute). Because it is a ratio of two extensive properties, concentration is an intensive property of the sample.

Molarity (M): Moles of species for consideration/total liters of solution

Volume %: Volume of species for consideration/total volume (same units for numerator and denominator)

Mass %: Mass of species for consideration/total mass (same units for numerator and denominator)

Although volume percent is widely used, the actual quantities can depend on the temperature and pressure of the species in the system, so volume percent varies with conditions. Mass percent, on the other hand, does not change with temperature and pressure.

For example, in the species $Mg(BH_4)_2$, the mass percent of Mg would be:

$$\text{mass \%(Mg)} = \text{mass of Mg/mass of } Mg(BH_4)_2 = 24.31 \text{ g}/54.01 \text{ g} = 0.45 \sim 45\%.$$

It is important to understand the terms solute, solvent, solution, homogeneous, and inhomogeneous when considering concentration.

In the United States and many other places in the world, the concentration of pure alcohol in an alcoholic beverage is called the proof. The proof of a drink is exactly equal to twice the volume percent, so pure ethanol is 200 proof. Bacardi 151 (151 proof) packs so much punch because it is 75% ethanol. The proofs of beer and wine are low, usually around 10 to 12 and 24 to 29, respectively, so volume percent is used more than the term *proof*. (Most beer is ~5% ethanol and wine is ~ 12 to 15%.)

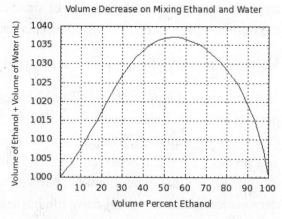

The alcohol content of beer is usually expressed as **volume percent** of ethanol (ABV, or alcohol by volume). This concentration unit refers to the volume of ethanol present in a given volume of solution. For example, one liter of a 20% by volume ethanol solution would be formed by mixing 200 mL of ethanol with enough water to make the final volume of one liter. It does not follow, though, that a 20% by volume solution of ethanol contains 80% by volume of water. When these liquids are mixed, the volumes are not additive. The mixing of the two solutions causes a change in the solution structure that decreases the volume. From the top of the graph at the right, one can see that 815 mL of water must be added to 200 mL of ethanol to bring the volume to one liter.

Conversion of concentration units from volume percent to mass percent requires knowledge of density of either the solute, the solution, or both. Molarity as a concentration unit refers to a volume of solution, so one need know only the density of the solute, ethanol, to convert volume percent to molarity. First convert the volume of ethanol in one liter (10 times its ABV) to its mass (assuming a density of 0.789 g/mL) and divide by the molar mass to get the number of moles in one liter.

Please note that the solutions used in this lab are combinations of 95% ethanol and water. Ingestion of ethanol at that concentration may damage to human mouths, throats, and internal organs. Accidental ingestion could be cause for emergency medical attention, so appropriate measures should be taken when handling the solutions to avoid inhalation, skin contact, or ingestion. Equally harmful, diethyl ether and sodium acetate are necessary for separation and require the same care when handling. Appropriate measures should be taken when handling the solutions to avoid inhalation, skin contact, or ingestion.

Efficiency of Separation

In the first term of general chemistry, we normally make calculations and predictions of reactions based on the premise that the reaction or process taking place proceeds to 100% completion. The reality in most reactions is that some of the process falls short of completion. The degree to which a reaction or process proceeds is called its efficiency. As mentioned in the section regarding the history process, the inability to completely recreate the process's more complex steps has led to a lower than 100% efficiency. As a result of many trials of the process below, it was determined that the efficiency of the process is ~85.0%. This value can be used to modify any results you may collect to determine the actual concentration of the unknown solution.

Procedure

Part I: Density of Ethanol

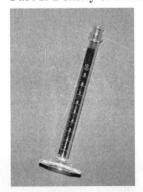

Weigh your 10 mL graduated cylinder to the nearest 0.0001 g or the smallest figure that the scales reads and record the weight in your notebook.

Obtain approximately 3.0 mL of pure ethanol in your graduated cylinder, and record the volume of ethanol to the nearest 0.01 mL.

Reweigh the graduated cylinder containing ethanol to the nearest 0.0001 g Dispose of the alcohol as directed and rinse the graduated cylinder.

Part II: Density of Water

Obtain approximately 3.0 mL of DI water in your previously weighed graduated cylinder, and record the volume of water to the nearest 0.01 mL.

Reweigh the graduated cylinder containing water to the nearest 0.0001 g.

Part III: Volume Percent of Alcohol

Weigh your clean, dry 25 mL graduated cylinder to the nearest 0.0001 g and record the weight in your notebook.

Add 25 mL of unknown to your preweighed graduated cylinder, touching the graduated cylinder only with a Kimwipe.
Record the exact volume of unknown to the nearest 0.01 mL. Weigh and record the mass of the graduated cylinder with unknown to the nearest 0.0001 g.

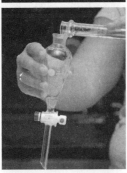

Carefully add the 25 mL of unknown to your separatory funnel; make sure the stopcock is CLOSED. Try to get as much of the liquid into the funnel as possible.

Using a weigh boat, carefully weigh out 1.0 g of NaCl. Record the exact mass in your lab notebook.

Add the salt to the separatory funnel and shake until the salt completely dissolves. Vent the funnel as demonstrated by your instructor and/or teaching/lab assistant about every 15 sec.
MAKE SURE THE FUNNEL IS POINTING AWAY FROM OTHER STUDENTS AND YOURSELF WHEN VENTING.

Add 10 mL of ethyl acetate (located in the properly ventilated hood) to your separatory funnel. Cap the funnel and shake and then vent, as shown by your TA prior to performing the experiment. Replace the cap and shake vigorously, venting periodically.
MAKE SURE THE FUNNEL IS POINTING AWAY FROM OTHER STUDENTS AND YOURSELF WHEN VENTING

Add 10 mL of diethyl ether (located in the properly ventilated hood) to your separatory funnel. Cap the funnel and shake and then vent, as shown by your TA prior to performing the experiment. Replace the cap and shake vigorously, venting periodically.
MAKE SURE THE FUNNEL IS POINTING AWAY FROM OTHER STUDENTS AND YOURSELF WHEN VENTING
Set the separatory funnel in a 400 mL beaker or a ring on your ring stand and allow the polar and nonpolar layers to separate. Two distinct layers should be visible.

Remove the cap and turn the stopcock to *slowly and carefully* drain the bottom yellow layer from your funnel into your preweighed 25 mL graduated cylinder. Try to get all of the yellow layer out of the funnel without losing any of the clear organic layer.

Dispose of the clear organic layer in the waste jar provided in the hood.

Record the volume and weight of the yellow liquid in your 25 mL graduated cylinder.

Dispose of the yellow liquid down the sink drain. Clean the graduated cylinder and separatory funnel thoroughly. Wash your hands well before leaving.

Report Contents and Questions

The purpose should consist of several well-constructed sentences stating what the experiment was designed to accomplish. Make sure to include statements about concepts and techniques explored in the experiments. The procedure section should reference the lab manual and include any changes made to the lab manual procedure during the lab.

The data section should include two tables. For Parts I and II, the table should include the following for both the ethanol and the water: **(a)** mass of graduated cylinder, **(b)** volume of liquid, **(c)** mass of liquid, and **(d)** density of liquid. For Part III, the table should include the following: **(a)** mass of graduated cylinder, **(b)** mass of graduated cylinder with solution, **(c)** mass of solution, **(d)** volume of solution, **(e)** density of solution, **(f)** mass of salt, **(g)** mass of water layer, **(h)** mass of alcohol, **(i)** volume of water layer, **(j)** volume of alcohol, **(k)** density of water layer, **(l)** percent alcohol by mass, **(m)** percent alcohol based on 85% efficiency, **(n)** percent alcohol by volume, and **(o)** proof of the alcohol. Also include any and all observations made during the extraction process.

A majority of the calculations required in the data section are straightforward and have been done in previous labs. The two new calculations are percent alcohol and percent alcohol based on the efficiency of the extraction. The percent alcohol by mass and volume is determined using the following formula (note that you need to convert the mass to volume using the densities of ethanol and water found in Part I):

$$\% \text{ alcohol} = \frac{\text{volume or mass of alcohol after extraction}}{\text{starting mass or volume of solution}} \times 100$$

Once you have determined the mass percent of alcohol, use the following formula to determine the percent alcohol based on the efficiency of the system.

$$\% \text{ alcohol based on efficiency} = \% \text{ alcohol} + (\% \text{ alcohol} \times ((100 - \% \text{ efficiency})/100)))$$

The calculation section should include example calculations of everything in the data section that was not collected during the actual experiment, including the following: **(a)** mass of liquid, **(b)** density of liquid, **(c)** mass of water layer, **(d)** mass of alcohol, **(e)** volume of alcohol, **(f)** percent alcohol based on mass, **(g)** percent alcohol based on volume, **(h)** percent alcohol based on 85% efficiency, and **(i)** proof of the alcohol solution.

The conclusion section should be several paragraphs in length and include the following information: a discussion for Part I, which includes the densities of the two liquids, and a complete discussion for Part II, which includes a comparison and explanation of the densities, volumes, and masses before and after the extraction. Give a detailed explanation of the percent alcohol values by mass and volume and how the 85% efficiency plays a role in determining the actual alcohol content of the unknown solution. Be sure to clearly state the percent alcohol content of your solution. Finally, discuss any errors.

Answer the following question:

1) Compare the experimental density of the unknown solution and the theoretical density (see graph below) of an alcohol solution with the mass percent you determined for your unknown. What is percent error between the two?

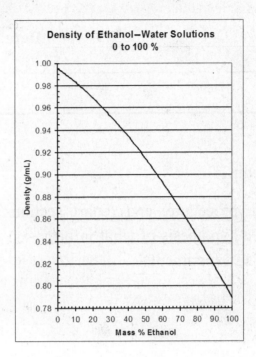

Experiment 5

Laboratory Preparation

Name: _____ **Date:** _____

Instructor: _____ **Sec. #:** _____

Show all work for full credit.

1) Read the background, procedure, and report sections of the lab experiment carefully and develop a hypothesis of what information you expect to gain from the completion of the lab experiment.

2) Create any and all tables you might need to collect data during the experiment and then transfer those tables into your lab manual for use during the lab.

Experiment 5

Prelaboratory Assignment

Name: _____ Date: _____

Instructor: _____ Sec. #: _____

Show all work for full credit.

1) You have a beer labeled as 7.3% alcohol. What is the molar concentration of ethanol?

2) It is more complicated to convert volume percent ethanol to mass percent ethanol because you must know the density of the solution. The graph below plots the density of aqueous ethanol solutions versus the volume percent ethanol.

What is the density of a 7.3% ethanol solution? (Note that the actual density of beer will be a bit different because of other components in the solution).

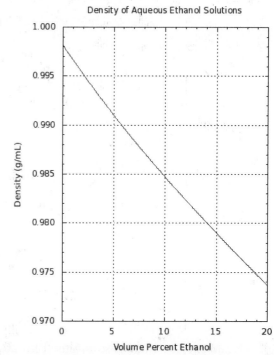

What is the mass of ethanol dissolved in one liter of a 7.3% ethanol solution?

What is the mass of water in one liter of a 7.3% ethanol solution?

What is the mass percent of ethanol (also known as ABW, alcohol by weight) in this solution?

3) In your laboratory experiment, you will determine the quantity of ethanol in an ethanol–water mixture indirectly by extracting the ethanol from the mixture and measuring the mass of the water remaining. Of necessity, such a procedure will give you the mass percent of ethanol in the mixture. To determine the proof of the alcohol, you must convert the mass percent ethanol to a volume percent, as the proof value in the United States is defined as twice the volume percent.

Your 50 mL graduated cylinder has a mass of 63.288 g. You measure approximately 31 mL of the unknown mixture in it and find that the total mass of the cylinder and the unknown is 92.244 g.

What is the mass of your unknown solution?

You place this unknown sample into your separatory funnel and add 1.026 g of NaCl and 10 mL each of ethyl ether and ethyl acetate. After extraction, you remove the bottom layer and find it has a mass of 22.057 g.

What mass of ethanol was removed by the extraction?

Complete extraction of the ethanol would require several extraction steps. To simplify the laboratory procedures and to cut down on errors and losses involved in transferring liquids, you will do only a single extraction. Assuming that prior experiments have shown that the single-step extraction removes only 70% of the ethanol, what is the mass of ethanol in your original unknown?

What is the mass percent ethanol in the original unknown?

Now, calculate the volume percent and the proof of the ethanol in the unknown solution based on this mass percent. Use the graph from the background information that plots the density of ethanol solutions from 20 to 60 mass percent to answer the following questions:

Given a density of 0.789 g/mL, what would the volume of the ethanol in the original unknown have been?

Using the density from the graph, calculate the volume of your original unknown solution. (The calculated volume should be more precise than the volume you could measure with your graduated cylinder.)

What would the volume percent of ethanol be in the original unknown solution?

What is the ethanol proof value of this unknown?

Experiment 6
Empirical Formula of a Compound

Introduction

When compounds are produced, the elements from which they are formed are always present in the same simple proportions. A chemical formula that expresses these simple proportions is known as an *empirical formula*. In this experiment, you will determine the empirical formula of an inorganic compound, magnesium oxide, a white substance that is formed when magnesium is burned in the presence of oxygen.

Background

At the beginning of your general chemistry course, you were introduced to a number of different laws. One of these was the law of conservation of mass, which states that the total mass of all reactants and products in a chemical reaction remains constant; that is, matter is neither created nor destroyed as a result of a process in which chemical changes take place. Another law was the law of definite proportions. This law is attributed to work done by Joseph Proust in the late eighteenth century. Based on numerous observations, Proust noted that a sample of any given compound, no matter how it was formed or where it was found, always has the *same composition*. Here, the term *same composition* is taken as the same mass proportions of all elements present in the compound. According to the law of definite proportions, any water found on Mars would have exactly the same mass ratio of hydrogen to oxygen as water on Earth.

Even though the law of definite proportions was originally expressed in terms of mass ratios, it also implies that fixed numbers of atoms are brought together to form compounds, with the molar mass of the element serving as a conversion factor. These numbers of elements are often expressed in a chemical formula for a compound. Chemical formulae (plural for *formula*) come in two varieties: empirical and molecular. An empirical formula expresses the general nature of the law of definite proportions in that it gives, as simple whole number ratios, the relative number of atoms of each kind in a compound. A molecular formula, on the other hand, provides the detailed number of each element in a compound.

Due to their generality, empirical formulae may actually represent two or more different chemical compounds. For instance, consider the molecular formulae for acetic acid (the active component of vinegar), CH_3COOH, and sucrose (a sugar), $C_6H_{12}O_6$. If you look closely, the simplest whole number ratio of carbon to hydrogen to oxygen for each compound is 1:2:1. This means that the empirical formula for both acetic acid and sucrose is CH_2O.

The empirical formula for a compound may be determined in a number of different ways. One such method is combustion analysis. In this example, a sample of known mass is placed in a combustion chamber through which a stream of pure oxygen is flowing. The sample is ignited, and the combustion products are swept out of the chamber through two materials that absorb the combustion gases, one material absorbing carbon dioxide and the other absorbing water vapor. Based on the increase in mass of the two absorbers, one can determine the empirical formula of the compound. The textbook presents numerous examples of combustion analysis.

This experiment will not be as complicated as a combustion analysis, but many of the same principles apply. This analysis starts with a known mass of magnesium that will be heated in the presence of air. The magnesium will react with nitrogen and oxygen in the air to form magnesium nitride and magnesium oxide, respectively. Addition of a small amount of water changes the magnesium nitride to magnesium oxide. The mass difference between the product and the pure magnesium sample will be the mass of oxygen that formed the oxide. Using the mass of magnesium before and the mass of magnesium oxide after combustion, we are able to find the grams of oxygen taken up by the solid during oxidation. By figuring out the number of moles of magnesium at the start, and the number of moles of oxygen taken up, it is possible to determine the empirical formula of magnesium oxide.

The Experiment

This experiment requires the use of an analytical balance. Make sure that crucibles and their covers are clean and dry. If a crucible or cover shows signs of a hairline crack, select another. The presence of such a crack may cause the crucible or cover to break when heated. The magnesium should be relatively clean. If it shows signs of corrosion, lightly sand the metal or select a new sample.

Notebook

At the start of the lab, it is necessary that the crucible, which will be the container for this high-temperature reaction, be "heated to constant mass." This means that the crucible is heated to drive off any water or other volatile material that is stuck to the crucible. The crucibles have been prefired once, so you can proceed by massing the crucible just after obtaining it; then proceed with the "second firing" and checking for constant mass (see below). You should record the initial mass in your notebook and record the mass after subsequent firings in your notebook. Sample tables for tracking the mass of the fired crucibles and the crucibles before and after the reaction are shown below. You will run this experiment twice to ensure reproducibility.

TABLE 6.1 Sample Data Table for Crucible Masses

Mass (g)	Crucible 1				Crucible 2			
	Bottom	Lid	Sum	Difference	Bottom	Lid	Sum	Difference
Initial				—				—
After First Firing								
After Second Firing								
After Third Firing								

TABLE 6.2 Sample Data Table for Reactant and Product Masses

		Trial 1		Trial 2	
		Lid	Crucible	Lid	Crucible
Empty Crucible (constant mass) (g)					
Crucible + Mg (g)					
Crucible + Oxide (g)					

Procedure

Initial Firing of Crucible and Lid

After obtaining the initial mass of a crucible and its lid, place these items (collectively known as the crucible from here on out) on a clay triangle. Start the Bunsen burner and adjust it to a well-defined, hot flame. Adjust the height of the iron ring so that the bottom of the crucible is just above the inner blue cone of the flame. Heat the crucible for several minutes, remove the flame, and allow the crucible to cool to room temperature. While waiting for the first crucible to cool, prepare your second trial and heat it up. Make sure to keep an eye on it. Remember to handle your crucibles with care, especially because of the heat. From this point, the crucible should be handled only with a pair of tongs. Mass the crucible to ± 0.0001 g on an analytical balance. Repeat this firing procedure one more time. If the mass between firings changes by more than 0.0100 g, additional firing may be necessary. Be careful to make sure that the crucible has cooled completely before placing it on the balance.

NOTES: Obtain the mass of the crucible bottom and lid separately. That way, if the lid is accidentally dropped, the experiment can continue with a substitute lid. Always handle the crucible with tongs! Fingerprints and other residue can change the mass of the crucible. It is a good idea to practice handling the crucibles with the tongs before starting the actual experiment.

Initial Magnesium Mass

Transfer about 0.3 g of magnesium into the bottom of the crucible. Use gloves, Kimwipes, or tongs for this operation to avoid leaving any finger oils on the metal surface. Fold or loosely coil the metal to fit into bottom of the crucible. Record the mass of the crucible and its contents to ± 0.0001 g using an analytical balance.

Conversion of Magnesium

Place the crucible on a clay triangle with its cover slightly opened to expose the magnesium sample to air. Gently heat the crucible with the upper part of the burner flame for a couple of minutes at the outset of the experiment. Increase the heat at this point by increasing the inner blue flame. If the magnesium starts to burn or white smoke is visible coming from the crucible, remove the flame and completely cover the crucible. After a few minutes, begin heating again. Once the metal no longer glows red and no more smoke is produced, the conversion reaction is complete. Apply strong heat to the crucible for several minutes more, making sure that its cover is slightly open. The crucible should glow a dull red color. Let the crucible cool off to room temperature.

Conversion of Magnesium Nitride

Magnesium nitride may form during the initial conversion process. This needs to be converted to magnesium oxide. After making sure your crucible is cool, use a glass stirring rod to gently push all of the crucible contents down to the bottom of the crucible. Be gentle, as some crucibles become fragile after heating. Next, add several (3 to 4) drops of deionized water to the crucible. Check for a faint smell of ammonia (NH_3). Cover the crucible and begin to heat it gently. Gradually increase the flame intensity until the crucible once again glows a dull red color. Continue heating with a hot flame for at least 5 min. Allow the crucible to cool to room temperature and then record its mass on an analytical balance to ± 0.0001 g. Repeat this firing procedure at least one more time until the mass of the crucible is within ± 0.02 g of the previous measurement.

The addition of water to the initial conversion product causes magnesium nitride (Mg_3N_2) to form magnesium hydroxide ($Mg(OH)_2$), releasing ammonia in the process. The hydroxide is converted to the oxide during the heating process by driving off water vapor. The difference between the mass of the empty crucible and the crucible/contents after this step is the mass of magnesium oxide.

Waste Disposal

Carefully scrape all of the solid material from your crucible into the container labeled "Magnesium Oxide Waste." As you dispose of the magnesium oxide waste, look to see if there is any metal left in the ash. Note this observation in your notebook.

Report Contents and Questions

Calculations

This section describes the calculations needed to successfully complete the laboratory assignment. Calculations should be attempted before leaving lab.

Initial Mass of Magnesium

The initial mass of magnesium, m_{Mg}, for each trial may be found by difference. Subtract the mass of the crucible, m_c, from the mass of the crucible with the magnesium in it, m_{c+Mg}:

$$m_{Mg} = m_{c+Mg} - m_c$$

Moles of Magnesium

To calculate the number of moles of magnesium used in the trial, n_{Mg}, divide m_{Mg} from above by the molar mass of magnesium, $24.3050 \text{ g·mol}^{-1}$:

$$n_{Mg} = m_{Mg} / (24.3050 \text{ g·mol}^{-1})$$

Mass of Oxygen Absorbed by the Conversion

The mass of oxygen absorbed by the conversion, m_o, is found by difference. Subtract m_{c+Mg} from the mass of the crucible and magnesium oxide, m_{c+Mg+o}

$$m_o = m_{c+Mg+o} - m_{c+Mg}$$

Moles of Oxygen

The moles of oxygen consumed, n_o, can be determined by dividing m_o by the molar mass of oxygen, $15.9994 \text{ g·mol}^{-1}$:

$$n_o = m_o / (15.9994 \text{ g·mol}^{-1})$$

Ratio of Magnesium to Oxygen

To determine the molar ratio of magnesium to oxygen, $Mg{:}O$, divide n_{Mg} by n_o:

$$Mg{:}O = n_{Mg}/n_o$$

Empirical Formula

Use the molar ratio of magnesium to oxygen to determine the empirical formula of magnesium oxide. It may be necessary to round to the nearest integer.

Answer the following questions:

1) Show how magnesium nitride is converted to magnesium oxide by writing two balanced chemical reactions that show the conversion process. Use the information at the end of the procedure to get started.

2) What would happen to the empirical formula of magnesium oxide if there were any unused magnesium metal left in the crucible at the end of the experiment? Would the ratio be too high in Mg or too high in O? Explain your answer.

3) Consider the possibility that your formula is Mg_2O_3. What does this imply regarding the charges of the ions? Do these charges agree with your expectations based on the periodic table? Explain.

4) Does your experimentally determined empirical formula agree with your expectation based on the periodic table? If not, suggest one or more reasons for the discrepancy.

Experiment 6
Laboratory Preparation

Name: _____ Date: _____

Instructor: _____ Sec. #: _____

Show all work for full credit.

1) Read the background, procedure, and report sections of the lab experiment carefully and develop a hypothesis of what information you expect to gain from the completion of the lab experiment.

2) Create any and all tables you might need to collect data during the experiment and then transfer those tables into your lab manual for use during the lab.

Experiment 6
Prelaboratory Assignment

Name: _____ Date: _____
Instructor: _____ Sec. #: _____

1) A 1.095 g sample of a compound containing only mercury and oxygen was heated until the oxygen was driven off as oxygen gas. The mass of the remaining sample was 1.053 g. Determine the empirical formula of this compound.

2) A student decided not to heat the crucible before adding his magnesium sample to it. Would you expect this decision to cause the molar ratio of magnesium to oxygen to be higher or lower than the actual value? Briefly explain your answer.

Experiment 7
Limiting and Excess Reagents

Introduction

This experiment is the culminating experiment for all the previous experiments involving stoichiometric explorations. In Experiment 4, we explored the 1:1 reaction between sodium bicarbonate and carbon dioxide, that is, for every *one* mole of sodium bicarbonate that reacted, *one* mole of carbon dioxide was formed. 1:1 reactions are very simple mathematically and thus a good starting point for stoichiometric investigations. In this experiment, we expand our knowledge by working with a reaction between copper(II) nitrate and potassium iodide, a reaction that does not have one-to-one stoichiometry.

In addition to stretching our ability to calculate product and reactant amounts using reaction stoichiometry, we will also be exploring the concept of limiting and excess reagents. It is not always obvious from gram amounts which reactant will run out first when a reaction occurs. Two factors complicate the prediction: (1) chemical equations are balanced in moles, not grams, and (2) moles are based on molecular weights. Thus, 100 g of O_2 is a lot fewer moles than 100 g of H_2.

The number of grams need to be converted to the number of moles, via the molar mass of each compound, accordingly, to get a clear understanding of why there are fewer oxygen molecule than hydrogen molecules as shown below:

$$100\,g\,O_2 \times \frac{1\,mol\,O_2}{32.0\,g\,O_2} = 3.13\,mol\,O_2 \qquad 100\,g\,H_2 \times \frac{1\,mol\,H_2}{2.02\,g\,H_2} = 49.5\,mol\,H_2$$

In addition to practicing stoichiometry, we will also learn about two new devices: the centrifuge and the desiccator.

 The centrifuge is a device that uses centrifugal force to separate solids (often called precipitates) from solution.

 A desiccator is a piece of laboratory equipment used to protect chemicals from moisture.

In a previous lab we mentioned substances called hydrates. A hydrate is a compound that has water molecules incorporated into its crystal structure. A desiccant is a hydrate that has been stripped of all its water molecules, making it very "hungry" for water. If a desiccant is placed in a sealed container, the container is then called a desiccator, since anything placed inside it will be stripped of its moisture by the desiccant.

Background

Many of the concepts used in this experiment, including limiting and excess reagents, moles, and theoretical yields, are covered extensively in your textbook. Since the treatment is extensive, another deep discussion here would be redundant, so we will just hit the highlights.

Limiting and Excess Reagents

The concept of limiting and excess reagents deals with how much product results when two or more reactants are mixed. This concept has very practical applications in the real world. For example, in the business of manufacturing a chemical product of some kind, it is important to do it as efficiently and with as little waste as possible. Generally, the most expensive reactant is the limiting reactant. This is also true in the laboratory, where you want to use only the amounts of reactants necessary to produce the largest amount of product, as anything more would be simply wasted.

Consider a simple reaction where one mole of reactant A reacts with one mole of reactant B:

$$A + B \rightarrow C$$

What happens if too much reactant A is added? When all of reactant B is used up, there will be some reactant A left over. In scientific terms, we say that reactant B is the limiting reagent. In other words, no matter how much A is added, no more product is made when B is consumed—reactant B limits how much product is obtained. On the other hand, reactant A is called the excess reagent because there is more than enough of reagent A and, in fact, there is leftover reagent A remaining at the end of the reaction.

In the reaction above, if we add more of the limiting reagent, reactant B, more of product C is formed. Why, you may ask? Well, reactant B limits how much product is obtained, so when more B is added, the reaction will resume until one of the reactants is used up again. Therefore, the final question is, "What is the most efficient mixture of these two reactants?" In fact, the best mixture consists of correctly matched amounts of reagent A and reagent B to allow for both reactants to be completely used. Note that unless the mole ratios are one to one, "correctly matched" does not mean "evenly matched."

Finding the Limiting Reagent

For practice, let's look at the following example and see if we can determine which reagent is limiting and which is in excess. Let's assume we have time-warped to the early 1970s and are all employed by NASA as rocket scientists. We are working with a fuel mixture composed of dinitrogen tetraoxide (N_2O_4) and hydrazine (N_2H_4). These two reagents react to produce nitrogen gas (N_2) and water vapor (H_2O), as shown below. We have been assigned to figure out which reactant is limiting when 1.40 kg of N_2H_4 and 2.80 kg of N_2O_4 are allowed to react.

$$N_2H_4(l) + N_2O_4(l) \rightarrow N_2(g) + H_2O(g)$$

The first step is to balance the equation:

$$2N_2H_4(l) + N_2O_4(l) \rightarrow 3N_2(g) + 4H_2O(g)$$

Next, we have to use stoichiometry to see how many moles of one of the products are produced based on the initial amounts of each reactant. Let's find the number of moles of H_2O.

Based on Moles of Hydrazine (N_2H_4)

$$1.40 \text{ kg } N_2H_4 \times \left(\frac{1000 \text{ g}}{1 \text{ kg}} \right) \times \left(\frac{1 \text{ mol } N_2H_4}{32.06 \text{ g } N_2H_4} \right) \times \left(\frac{4 \text{ mol } H_2O}{2 \text{ mol } N_2H_4} \right) = 87.3 \text{ mol } H_2O$$

Based on Moles of Dinitrogen Tetraoxide (N_2O_4)

$$2.80 \text{ kg } N_2O_4 \times \left(\frac{1000 \text{ g}}{1 \text{ kg}} \right) \times \left(\frac{1 \text{ mol } N_2O_4}{92.02 \text{ g } N_2O_4} \right) \times \left(\frac{4 \text{ mol } H_2O}{1 \text{ mol } N_2O_4} \right) = 122 \text{ mol } H_2O$$

Since hydrazine produces fewer moles of water, it must be the limiting reagent!

As you can now see, it is very important to calculate exactly how much product can be made based on the stoichiometry of each and every reactant. Another point to note is that in any reaction, there is always a portion of the product that is lost to human error, incomplete mixing of reagents, and so on. Therefore, it is important in any chemical preparation to take this loss into account so that enough product is made.

Precipitation

In this experiment, the reaction of the two solutions, copper(II) nitrate, $Cu(NO_3)_2$, and potassium iodide, KI, forms a precipitate. A precipitate is a solid formed during a reaction between two aqueous compounds. (This is the same word meteorologists use when talking about falling rain or snow.) To better separate the solid and liquid phases, test tubes containing precipitates are placed in a centrifuge. By spinning them around at an extremely high velocity, the centrifuge forces the heavier precipitate to the bottom of the test tube. After centrifugation, a solution that was cloudy becomes clear with a solid accumulated at the bottom of the tube.

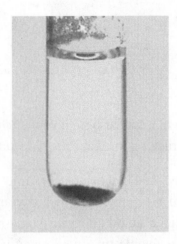

The clear liquid found above the precipitate is called the supernatant. In this experiment, we will test the supernatant to see which reactant remains after the reaction is complete. The reactant that is present in the supernatant when the reaction stops is obviously in excess, while the reactant not detected in the supernatant was completely used and is our limiting reagent. This can be verified by adding more of the two solutions we are studying. Using our earlier example, if more solution A is added and additional precipitate forms, then reagent A is the limiting reagent. Similarly, if more solution B is added and additional precipitate forms, then reagent B is the limiting reagent.

The Reaction in the Experiment

Specifically, this experiment involves mixing aqueous solutions of $Cu(NO_3)_2$ and KI, which will react as shown:

$$2Cu(NO_3)_2(aq) + 4KI(aq) \rightarrow 2CuI(s)\downarrow + I_2(aq) + 4KNO_3(aq)$$

Since we are dealing with solutions, the concentration is expressed in molarity, a term used in our conservation of matter experiment (Experiment 3). Recall that molarity (M) is defined as the number of moles of solute in one liter of solution. If we know the volume of solution, we can then calculate the number of moles present by multiplying the volume in liters by the molarity of the solution:

$$L_{solution} \times \frac{moles}{L_{solution}} = moles$$

A reaction between 0.50 M $Cu(NO_3)_2$ and 0.50 M KI now has a little more meaning to us. In the experiment, each student will be assigned different volumes of reactants to investigate. Based on the volume of reactants assigned, a theoretical amount of product can be calculated. After the solutions react, the precipitate is recovered and its mass determined. The final step in the process is to evaluate the overall efficiency of the reaction by reporting the percent yield of the reaction based on the experimental results.

Procedure

SAFETY NOTES: The supernatant in this experiment contains I_2. The I_2 will leave a yellow stain on your skin if you come in contact with it. I_2 will also stain your clothes, so be careful. **If you know you are allergic to iodine, notify your instructor immediately so that further precautions may be taken.**

GENERAL INSTRUCTIONS: To make the most of time in the lab, the class will pool all results. Be sure to fill in the table of the class results when you are finished with your experiment. Since some students will finish earlier than others, bring something to read or study along with you to class so that you will not waste valuable time while waiting for the rest of the class to finish. You must have these results to complete the lab report.

NOTE: The pictures below are for demonstration purposes only and may show improper technique with respect to handling of the test tubes in order to produce a clearer pictures. You, however, should handle weighed test tubes only with Kimwipes or clamps.

Part I: The Precipitation Reaction

Collect six small test tubes (clean and dry) as well as clamps or tongs. Label the test tubes 1, 1a, 1b, 2, 2a, and 2b. The test tubes marked 1 and 2 should be Pyrex because they will be heated later in the experiment. Wipe off any fingerprints or other smudges from tubes 1 and 2, and then weigh them carefully to the nearest 0.0001 g. Do not touch them with your fingers from now on. Test tubes 1a, 1b, 2a, and 2b will not be weighed and can be touched as needed.

Each student in the lab section will be assigned certain volumes of the 0.50 M $Cu(NO_3)_2$ and 0.50 M KI. Tubes 1 and 2 will have identical mixtures so that if any error is made, there is a duplicate. Add the assigned volume of $Cu(NO_3)_2$ to the assigned volume of KI in tubes 1 and 2. Be sure to record any observations of the reaction in your lab notebook.

Holding the top of the test tube gently but firmly between the thumb and forefinger of one hand, tap the bottom of the test tube gently with the forefinger of your other hand to thoroughly mix the chemicals. Set the tubes in the test tube rack for 5 min to allow the reaction to go to completion. Remember not to touch test tubes 1 and 2 with your hands. Use a clamp, tongs, or a Kimwipe.

Centrifuge the test tubes to collect the solid. (Your instructor will demonstrate the proper use of the centrifuge.) Place the two test tubes directly opposite each other. Make sure they are seated properly in the holders. Each centrifuge can accommodate more than two test tubes, and the experiment will go much faster if several people use the centrifuge at the same time. Make sure you can identify your test tubes, both by the labels and by the position in the centrifuge! Let the test tubes spin for 5 min. Turn off the centrifuge and wait until the rotor has completely stopped before attempting to remove your test tubes.

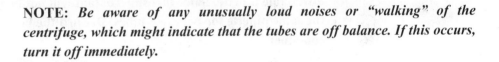

NOTE: *Be aware of any unusually loud noises or "walking" of the centrifuge, which might indicate that the tubes are off balance. If this occurs, turn it off immediately.*

Remove the test tubes, being careful not to touch them with your fingers. They should contain a clear brown-green colored liquid (supernatant) and a gray solid at the bottom (precipitate). Put the test tubes in the test tube rack.

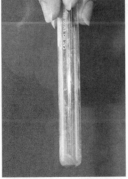

The next step is to draw off the supernatant, leaving the precipitate in the test tube. It is not essential that every last drop of supernatant is removed from the test tube, but it is important that the precipitate be undisturbed, so follow these directions carefully. Obtain two disposable pipettes. Squeeze the air out of the pipette before putting it into the liquid and then carefully draw up the supernatant.

Split the supernatant from test tube 1 between the test tubes labeled 1a and 1b. Discard the pipette as instructed. Use the second pipette for test tube 2 and follow the same procedure.

Test tubes 1a, 1b, 2a, and 2b contain an excess of either $Cu(NO_3)_2$ or KI. The supernatants will be tested later, so don't throw them away. Test tubes 1 and 2 contain the precipitates. This precipitate is wet and must be dried before it can be weighed. Any unreacted reagents that may be trapped must also be removed.

Wash the precipitate by adding approximately 3 mL methanol to each test tube. It is important that the same amount be added to both test tubes. Gently shake the tubes so that the precipitate starts to cloud the solution. Remember not to touch the tubes with your fingers. Also, be careful not to spill any methanol on the outside of the test tubes because it will dissolve the marker.

Spin the test tubes in the centrifuge for 5 min. The supernatant should be clear once again.

Without disturbing the precipitate, remove the supernatant with a disposable pipette. Since this supernatant will not be tested, you can use the same pipette for both test tubes. Discard the liquid in the proper container.

To dry the precipitate, place test tubes 1 and 2 in a 100 mL beaker. Write your name on a label or a piece of paper and place it on the inside of the beaker. Put the beaker and test tubes into the drying oven for at least 20 min. If there is more time left in the period, leave them in longer, keeping in mind that the test tubes must be cooled before weighing.

NOTE: *Methanol, rather than water, is used to wash the precipitate because methanol is much more volatile. This means that it evaporates very quickly, making the precipitates dry much faster. Methanol is also flammable, so no flames in the lab today!*

Part II: Testing the Supernatants

While the test tubes are in the oven, test the supernatants. Add 1 mL of the $Cu(NO_3)_2$ solution to test tubes 1a and 2a. Record any observations.

Now add 1 mL of the KI solution to test tubes 1b and 2b and record observations.

Part III: Weighing the Precipitates

After the test tubes have been in the oven at least 20 min, remove them and place the test tubes in a desiccator to cool for at least 10 min. Remember not to touch the test tubes with your fingers (the beaker will be hot anyway).

Weigh each test tube to the nearest 0.0001 g.

Calculate the mass of precipitate for each test tube. Give your instructor the results to post for the whole class. Make sure you have a copy of everyone's results before leaving the lab.

Be sure that precipitates, supernates, and methanol wash are all disposed of in the proper containers.

Report Contents and Questions

The purpose should be several sentences stating the reasons for doing this experiment and the criteria used for evaluating success. It should include both concepts and laboratory techniques. The procedure section should cite the lab manual and note any changes that were made to the procedure in the lab manual while performing the experiment.

The data section should include your own personal data, as well as the table below filled in with class data. For your personal data, include all observations, the masses, and the supernatant test results, using a "+" (more precipitate formed) or "–" (more precipitate did not form). Your personal data should be in a data table similar to that of the class data.

Class Data Table

	Vol. $Cu(NO_3)_2$ (0.50 M)	Vol. KI (0.50 M)	Masses of ppt. in test tubes	$Cu(NO_3)_2$ Test	KI Test
A	1.0 mL	6.0 mL			
	1.0 mL	6.0 mL			
	1.0 mL	6.0 mL			
	1.0 mL	6.0 mL			
B	2.0 mL	5.0 mL			
	2.0 mL	5.0 mL			
	2.0 mL	5.0 mL			
	2.0 mL	5.0 mL			
C	3.0 mL	4.0 mL			
	3.0 mL	4.0 mL			
	3.0 mL	4.0 mL		—	+
	3.0 mL	4.0 mL			
D	4.0 mL	3.0 mL			
	4.0 mL	3.0 mL			
	4.0 mL	3.0 mL			
	4.0 mL	3.0 mL			
E	5.0 mL	2.0 mL			
	5.0 mL	2.0 mL			
	5.0 mL	2.0 mL			
	5.0 mL	2.0 mL			
F	6.0 mL	1.0 mL			
	6.0 mL	1.0 mL			
	6.0 mL	1.0 mL			
	6.0 mL	1.0 mL			

The balanced chemical equation should be displayed above the data table. The following should be calculated for each group and placed in the class data table: **(a)** average moles of product (CuI), **(b)** average moles of $Cu(NO_3)_2$, **(c)** average moles of KI, **(d)** theoretical moles of product based on $Cu(NO_3)_2$, **(e)** theoretical moles of product based on KI, **(f)** which reactant is the limiting reagent, and **(g)** percent error in moles of CuI.

The calculation section should have sample calculations for the following: **(a)** moles of $Cu(NO_3)_2$, **(b)** moles of KI, **(c)** moles of product (CuI), **(d)** average number of moles observed, **e)** theoretical moles of product based on $Cu(NO_3)_2$, **(f)** theoretical moles of product based on KI, **(g)** determination of the limiting reagent, and **(h)** percent error of moles of CuI.

Also include a graph of the number of moles of CuI expected versus the number of moles of KI added. All graphing guidelines apply here.

The conclusion should be in paragraph form and include a summary of the information determined by performing the experiment. This section should include the findings for each set of solutions. For solutions A through F, include the test results, experimental product yield, theoretical yield, limiting reagent, and percent error. Compare the results of your predicted limiting reactant with the results of the tests on the supernatant solutions. Be sure to discuss the variation in the different solutions and what caused the differences. Also discuss what information can be determined from the graph. Finally, discuss all sources of error and how they might have affected the results.

Answer the following questions:

1) Explain why the theoretical yield of a reaction is seldom accomplished in the laboratory.

Experiment 7
Laboratory Preparation

Name: _____ Date: _____

Instructor: _____ Sec. #: _____

Show all work for full credit.

1) Read the background, procedure, and report sections of the lab experiment carefully and develop a hypothesis of what information you expect to gain from the completion of the lab experiment.

2) Create any and all tables you might need to collect data during the experiment and then transfer those tables into your lab manual for use during the lab.

Experiment 7
Prelaboratory Assignment

Name: _____ Date: _____

Instructor: _____ Sec. #: _____

Show all work for full credit.

1) Your laboratory experiment on limiting and excess reagents involves the reaction between aqueous solutions of $Cu(NO_3)_2$ and KI to form CuI, KNO_3, and I_2. The CuI precipitates from the reaction as a solid, and the other products (KNO_3 and I_2) remain in aqueous solution. In the lab, you will vary the relative amounts of each reactant, determining the amount of solid product in each case. In this prelab assignment, you will explore some features of this reaction.

Write a balanced molecular equation for the overall reaction, using the smallest whole number coefficients and including the states of matter for each reactant and product.

This reaction is an oxidation–reduction (redox) reaction in which one reactant loses electrons (is oxidized) and another reactant gains electrons (is reduced). To recognize a redox reaction, you need to determine the oxidation states of all the elements in both the reactants and products. If any element changes its oxidation state, then it is a redox reaction. An increase in oxidation state is an oxidation; a decrease in oxidation state is a reduction. (NOTE: If an element is a reactant or product, converted to or formed from a compound, the reaction must be a redox reaction.)

Indicate the oxidation state of the elements in each of the following compounds:

Reactants   Products

$Cu(NO_3)_2$ KI   CuI KNO_3 I_2

Cu_____   K_____   Cu _____   K_____   I_____

N _____   I _____   I _____   N _____

O _____   O _____

Which reactant is oxidized, and which is reduced?

97

2) Calculating molarity:

What is the molar mass of CuI?

What quantity of CuI is present in 226 mL of a 0.20 M solution?

What is the molar mass of $Cu(NO_3)_2$?

What mass of $Cu(NO_3)_2$ is present in 151 mL of a 0.60 M solution?

What is the molar mass of KI?

What mass of KI is needed to add to a 1000 mL volumetric flask to produce a solution that is 0.40 M?

What is the molar mass of I_2?

What is the molarity of a solution produced by dissolving 41 g of I_2 in a 500 mL volumetric flask?

3) Calculate limiting reagent:

You mix 6.0 mL of 0.50 M $Cu(NO_3)_2$ with 1.0 mL of 0.50 M KI and collect and dry the CuI precipitate. Calculate the following:

Moles of $Cu(NO_3)_2$ used:

Moles of KI used:

Moles of CuI precipitate expected from $Cu(NO_3)_2$:

Moles of CuI precipitate expected from KI:

What mass of CuI do you expect to obtain from this reaction?

If the actual mass of the precipitate you recover is 0.038 g, what is the percent recovery of the precipitate?

Experiment 8

Redox Reactions in Voltaic Cells:
Construction of a Potential Series

Introduction

Have you ever wondered how a battery works? We use them every day for our cell phones, our cars, timers, alarm clocks and even some important things like pacemakers. But have you ever really thought about how or why they work? What's inside of them? For one second, just imagine that there was no power, no electricity coming from the plug in the wall, but you desperately needed power to make something work because your life depended on it. What would you do? What (as a chemist) could you do?

Electrochemistry is the study of electrochemical (oxidative and reductive) reactions. These are reactions that create a flow of electrons from one place to another. A flow of electrons is electricity. A battery is a small holding cell of electrochemical energy. In each battery, there are two reservoirs of chemicals that allow a transfer of electrons if connected by a conductive material. To use this flow of electrons (electricity) to power our devices, we simply need to wire them into the loop:

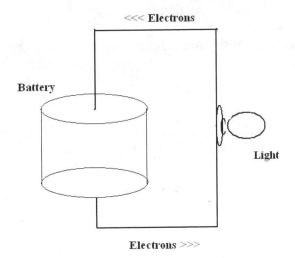

As a chemist, you should know the basic design of a common battery and also the chemistry behind why it works. In this lab, we are going to create some small voltaic cells out of various chemicals and measure their potentials (i.e., how strong the flow of electrons is from one metal to another). The greater the potential difference (the ease with which the metals gain electrons) between two metals, the stronger the electron flow from one to the other; the stronger the electron flow, the more electricity is generated. A lot of research has been devoted to the production of a better battery. The lithium batteries that power our laptops are the result of some of that research and are approximately twice as strong as a normal zinc-carbon battery.

Developing a potential series of metals is therefore a valuable tool for chemists, allowing them to select metals/chemicals that will produce the most power.

In addition to learning about redox reactions, potentials, and the chemistry behind a battery, you will also learn to use a new device called a voltmeter or multimeter. We will use a voltmeter or multimeter in the lab to measure the potentials of the cells that we construct.

Background

Redox

In both the conservation of matter and limiting and excess reagents experiments, you were given a brief introduction to redox reactions. Recall that redox reactions are also known as oxidation–reduction reactions. In particular, *oxidation* and *reduction* are the terms used to describe the transfer of electrons in a chemical reaction. Atoms or ions that accept electrons in a reaction have undergone reduction, while species that lose electrons undergo oxidation. These two definitions can be remembered by either of these two mnemonics:

1) LEO the lion goes GER—Loses Electrons, Oxidized (LEO); Gains Electrons, Reduced (GER)

2) OIL RIG—Oxidation Is Loss; Reduction Is Gain

When an oxidation reaction and a reduction reaction are paired together, electrons can flow from the oxidized species (losing electrons) to the reduced species (gaining electrons). The reduced species is often referred to as the oxidizing agent, or oxidant, as it is taking electrons away from another atom or ion, causing oxidation. Conversely, the species being oxidized is referred to as the reducing agent or reductant because it is donating electrons, thus causing reduction. The oxidation or reduction of an atom or ion can be followed by observing the species' oxidation number on the reactant side (left) and seeing if it has changed on the product side (right). The rules for assigning oxidation numbers are available in your textbook.

Electricity and Voltaic Cells

In general, the transfer of electrons from one place to another is referred to as electricity. In fact, the electricity that you use to power your phone is simply the flow of electrons along a metal wire. Another way to produce electricity is through the spontaneous redox reactions that occur in voltaic cells. The oxidation and reduction processes are separated so that the transfer of electrons occurs through an external wire. These "separated" parts of the cell are called half-cells, one for oxidation and one for reduction.

The construction of a voltaic cell takes advantage of the spontaneous nature of a redox reaction. For example, let's look at the following reaction in which zinc metal is oxidized and copper ion is reduced.

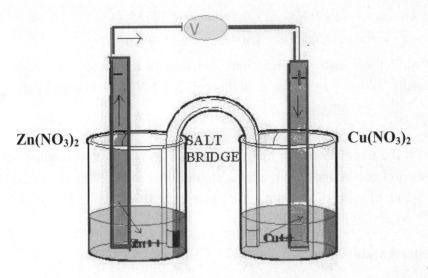

If we were to construct the characteristic voltaic cell, we would need 1.0 M solutions of copper(II) nitrate, $Cu(NO_3)_2$, and zinc nitrate, $Zn(NO_3)_2$. Additionally, we would have to suspend a piece of copper metal in the container holding the $Cu(NO_3)_2$ solution and a piece of zinc in the $Zn(NO_3)_2$, as shown in the diagram above. Both metals serve as electrodes, the conductor used to allow contact with the ions in solution. The solutions are then connected by a third solution, which is called the salt bridge.

The piece of zinc metal is the negative electrode, referred to as the anode, while the piece of copper metal is the positive electrode, called the cathode. Finally, when the circuit is completed by connecting the two electrodes with an external metal wire, usually a voltmeter, the flow of electrons can be observed.

When a voltaic cell is constructed correctly, oxidation always occurs at the anode, while reduction always occurs at the cathode. A little hint to help you differentiate these two components is that oxidation and anode both start with vowels, while cathode and reduction both start with consonants. Thus, the flow of electrons should always have a positive potential value because they flow spontaneously from negative (anode) to positive (cathode). In case you do obtain a negative potential value, which indicates that your electrons are flowing in the opposite (nonspontaneous) direction, all you have to do is switch the electrode connections on the voltmeter.

In this example, the electrons flow away from the zinc and toward the copper. This means that the copper ions are being reduced to copper atoms (gaining electrons) while the zinc atoms are being oxidized to zinc ions (losing electrons). Furthermore, since copper is being reduced, we say that it is the stronger oxidizing agent, and, by default, zinc has to be the stronger reducing agent. The direction of the flow of electrons depends on the potential of each metal ion.

Let's look at the reactions taking place at the individual electrodes. Recalling that oxidation occurs at the anode and reduction occurs at the cathode, we can break down our reaction into two half-reactions:

$$\text{Anode:} \quad Zn^0(s) \rightarrow Zn^{2+}(aq) + 2e^-$$
$$\text{Cathode:} \quad Cu^{2+}(aq) + 2e^- \rightarrow Cu^0(s)$$

We can then "add" these two half reactions together to generate the overall reaction given previously in this discussion. The potential of this cell is 1.1 V as measured by the voltmeter.

Standard Cells

A cell in which all the materials are in their standard states is called a standard cell. The standard states of elements and compounds are the most stable form found naturally at 25 °C. For gaseous substances, the standard state is 1.00 atm of pressure and 0 °C, while for solutions, the standard state is 1.00 M, 1.00 atm of pressure, and 25 °C.

Voltaic Cell Representation

Rather than writing out long chemical equations to represent redox reactions occurring in voltaic cells, a short hand form has been devised using the following rules.

1)   The negative electrode (anode) is written on the left side, and the positive electrode (cathode) on the right side.

$$\textbf{Zn} \qquad\qquad \textbf{Cu}$$

2)   All materials involved in the cell are represented with symbols and formulas, including the physical state.

$$\textbf{Zn}^0_{(s)}\ \textbf{Zn}^{2+}_{(aq)} \qquad\qquad \textbf{Cu}^{2+}_{(aq)}\ \textbf{Cu}^0_{(s)}$$

3)   Direct contact is indicated by a single vertical line.

$$\textbf{Zn}^0_{(s)}|\textbf{Zn}^{2+}_{(aq)} \qquad\qquad \textbf{Cu}^{2+}_{(aq)}|\textbf{Cu}^0_{(s)}$$

4)   Indirect contact through a salt bridge is indicated by double vertical lines ‖.

$$\textbf{Zn}^0_{(s)}|\textbf{Zn}^{2+}_{(aq)}\,\|\textbf{Cu}^{2+}_{(aq)}|\textbf{Cu}^0_{(s)}$$

5)   The concentrations of the electrolytes are given in parentheses.

$$\textbf{Zn}^0_{(s)}|\textbf{Zn}^{2+}_{(aq)}\textbf{(1.0 M)}\,\|\textbf{Cu}^{2+}_{(aq)}\textbf{(1.0 M)}|\textbf{Cu}^0_{(s)}$$

Calculations of a Cell's Potential

A term not mentioned in the previous sections is the *overall cell potential* (E^0_{cell}), which is simply the sum of the standard reduction potential (E^0_{red}) and the standard oxidation potential (E^0_{ox}). Mathematically, the overall cell potential is expressed as:

$$E^0_{cell} = E^0_{red} + E^0_{ox}$$

Using the example shown in our diagram above, let's make a reasonable hypothesis about the overall potential for this cell. In the figure, we see that Zn is being oxidized at the anode, while Cu^{2+} is being reduced at the cathode, as shown in the half-reactions provided below:

Anode: $\quad Zn^0(s) \rightarrow Zn^{2+}(aq) + 2e^-$

Cathode: $\quad Cu^{2+}(aq) + 2e^- \rightarrow Cu^0(s)$

If you look in Appendix D, you will find a list of reduction potentials for a great many elements, ions, and compounds. Because these are reduction potentials, to determine the oxidation potential, we must reverse the sign of the potential value. Remember that oxidation (the loss of electrons) is simply the reverse of reduction (the addition of electrons) for an element, ion, or compound. If we use Appendix D to look up the potentials for copper and zinc, we find that the E^0_{red} for Cu^{2+} is +0.34 V, while the E^0_{red} for Zn^{2+} is –0.76 V. To calculate the E^0_{ox} for Zn, we must reverse the sign, +0.76 V. Then we can calculate our predicted overall cell potential:

$$E^0_{cell} = 0.34 \text{ V} + 0.76 \text{ V} = +1.10 \text{ V}$$

However, when you construct this cell in the experiment, the reading from your voltmeter may not be exactly +1.10 V. This is because you are probably not working at completely standard conditions. However, if the ratio of concentrations of the corresponding salts is 1:1, any differences in the readings could be explained by other factors such as resistance (lack of good electron flow) in the salt bridge or lack of calibration of the voltmeter.

Procedure

SAFETY NOTES: Use the same precautions required when working with any chemical solution.

GENERAL INSTRUCTIONS: Each student should complete a series of measurements for three metals and share the results with the class as directed by your instructor.

Part I: Building the Cells

Obtain four styrofoam cups. Cut the tops off the cups so that the part of the cup that remains is ~5 cm deep.

Obtain a sheet of white cardstock or cardboard and tape the cups to the paper and label what is in each cup accordingly as shown in the figure to the left.

Fill the cup in the middle about half full with 0.5 M NH₄NO₃.

Fill each of the other three cups half full with one of your assigned metal nitrate solutions. Make sure to label the paper to the side of the cup with the name and concentration of the contents of the cup.

Obtain three filter papers and fold them into 1 cm strips.

Bend one of the strips into a U shape. Place one end of the U into the center cup, make sure that it contacts the solution, and then place the other end into any of the other three cups. The filter paper will act as a salt bridge between the three solutions you are measuring. You may need to presoak the filter paper strips in NH₄NO₃ to make sure the salt bridge is functioning properly. Repeat until there is a salt bridge between the center cup and all three of the other cups.

Part II: Measure the Potentials

Turn on the voltmeter. Set the voltmeter to read volts (V dc).

Obtain a piece of each of your assigned metals. Use sandpaper to "shine" the surfaces of the metal pieces.

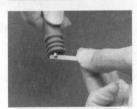

Attach the positive end (red) of the voltmeter to a piece of metal and place it in the correct (matching) nitrate solution so that the metal piece is partially submerged. (DO NOT SUBMERGE THE ALIGATOR CLIPS IN THE SOLUTION.) Use tape to adhere the clips and metal in place if necessary. It is easiest if you simply hold both clips in place.

Attach the negative end (black) of the voltmeter to another of your assigned pieces of metal and place it in the correct (matching) nitrate solution so that the metal piece is partially submerged. (DO NOT SUBMERGE THE ALLIGATOR CLIPS IN THE SOLUTION.) Use tape to adhere the clips and metal in place if necessary. It is easiest if you simply hold both clips in place.

Measure the potential and record it in your lab notebook. If the potential is negative, switch the alligator clips between the metals.

Repeat the potential measurements until a potential is established for each metal combination. Make sure to record your findings in your notebook.

Report Contents and Questions

The purpose should be several well-constructed sentences dealing with the major teaching points of this experiment. Discuss both conceptual and technique-based topics and your criteria for determining success. The procedure section should cite the lab manual and also note any changes made to the procedure in the lab manual. This section should also state which three metals you used for the experiment.

The data section for this experiment should include a data table similar to the one below:

Cell Representation	Anode Half-Reaction	E^0_{anode}	Cathode Half-Reaction	$E^0_{cathode}$	Overall Reaction	E_{cell} Theo.	E_{cell} Exp.	% error
$Fe^0|Fe^{2+}|| Cu^{2+}|Cu^0$	Fe^0 to $Fe^{2+} + 2e^-$		$Cu^{2+} + 2e^-$ to Cu^0		$Fe^0 + Cu^{2+}$ to $Fe^{2+} + Cu^0$			
Second cell								
Third ell								
Fourth cell								
...n^{th} cell								

An example of a voltaic cell and half-reaction are included in this data table. Use this example to complete the cells you did. The E_{cell} theoretical (E_{cell} Theo.) can be determined using the half-cell potential table found in your textbook. Simply find the half-reactions of your cell and use the following formula: $E_{cell} = E_{cathode} + E_{anode}$. The E_{cell} experimental value is the voltage you measured with your voltmeter.

The calculation section should include sample calculations for determining E_{cell} theoretical.

The conclusion section should be several well-developed paragraphs that cover the following information: **(a)** a potential series ranking all of the metals tested by the class; these metals should be arranged from strongest to weakest oxidizing agent and **(b)** a potential series ranking the metal *ions* from the strongest to the weakest oxidizing agent. Be sure to explain the reasoning that led from your data to your conclusions. A discussion on the relationship of the weakest reducing metal to the strongest oxidizing metal should also be included. Another discussion should be included on the comparison of the E_{cell} theoretical and E_{cell} experimental values along with the percent error values. Finally, discuss any errors that might have led to the discrepancies in the E_{cell} values and any changes that should be made in the experiment's procedure to compensate for these errors.

Answer the following question:

1) What is a reactivity series, and how is it related to this experiment?

Experiment 8
Laboratory Preparation

Name: _____ Date: _____

Instructor: _____ Sec. #: _____

Show all work for full credit.

1) Read the background, procedure, and report sections of the lab experiment carefully and develop a hypothesis of what information you expect to gain from the completion of the lab experiment.

2) Create any and all tables you might need to collect data during the experiment and then transfer those tables into your lab manual for use during the lab.

Experiment 8
Prelaboratory Assignment

Name: _____ Date: _____

Instructor: _____ Sec. #: _____

Show all work for full credit.

The following is a diagram of a voltaic cell.

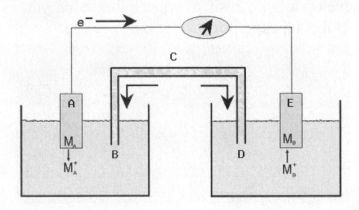

1) Identify the parts of the cell shown above by selecting the appropriate letter from the figure.

_____This is the positive electrode.

_____Reduction occurs at this electrode.

_____This electrode is the cathode.

_____Anions depart from the salt bridge here.

_____Oxidation occurs at this electrode.

2) Write the shorthand notation for the following voltaic cell reaction:
$$Fe(s) + Cu^{2+}(aq) \rightarrow Fe^{2+}(aq) + Cu(s)$$

Write the reaction represented by the following voltaic cell notation:
$$Zn(s)|Zn^{2+} (1\ M)\ ||Cu^{2+} (1\ M)\ |Cu(s)$$

3) The list below shows the reduction potentials for 10 different metal ions.

Calculate ΔE for the redox reaction: $Co(s) + Pb^{2+}(aq) \rightarrow Co^{2+}(aq) + Pb(s)$
Will this reaction occur spontaneously?

Calculate ΔE for the redox reaction: $Mn(s) + Fe^{2+}(aq) \rightarrow Mn^{2+}(aq) + Fe(s)$
Will this reaction occur spontaneously?

Half-Reaction	$E^0(V)$
$Cu^{2+}(aq) + 2e^- \rightarrow Cu(s)$	+0.34
$Pb^{2+}(aq) + 2e^- \rightarrow Pb(s)$	–0.13
$Sn^{2+}(aq) + 2e^- \rightarrow Sn(s)$	–0.14
$Ni^{2+}(aq) + 2e^- \rightarrow Ni(s)$	–0.25
$Co^{2+}(aq) + 2e^- \rightarrow Co(s)$	–0.28
$Cd^{2+}(aq) + 2e^- \rightarrow Cd(s)$	–0.40
$Fe^{2+}(aq) + 2e^- \rightarrow Fe(s)$	–0.44
$Zn^{2+}(aq) + 2e^- \rightarrow Zn(s)$	–0.76
$Mn^{2+}(aq) + 2e^- \rightarrow Mn(s)$	–1.18
$Mg^{2+}(aq) + 2e^- \rightarrow Mg(s)$	–2.37

Experiment 9
Reactions in Aqueous Solutions: Strong Acids and Bases

Introduction

Strong acids and bases are a part of our daily lives, even though we seldom think about them. The gastric acid in your stomach is a strong (pH 2 to 4) hydrochloric acid solution that hydrolyzes (breaks) the bonds in the carbohydrates and proteins that make up your food. The drain cleaner that unclogs your sink contains concentrated sodium hydroxide, which converts grease or oils in the clog into soap, which is then soluble in water and can be washed away. In this experiment, we will test the relationship between pH and acid or base concentration. We will also discuss how strong acids and bases react to neutralize each other, forming predictable products of salt and water.

Background

Acid and Base Strength

The strength of an acid or base is determined by its ability to dissociate in water. An acid that completely dissociates to form its component ions H^+ and A^- when placed in water is a strong acid. The same can be said for a base that completely dissociates into B^+ and OH^-. Weak acids and bases do not completely dissociate into their component ions and thus add far fewer H^+ or OH^- ions to the solution.

Acid and base strengths are characterized quantitatively by a value called the K_a and K_b, (acid/base dissociation constants) respectively. These dissociation constants are calculated as a ratio of the amount of acid or base that is separated into ions compared to the amount of acid or base that is still together in acid or base form.

$$K_a = \frac{[H^+][A^-]}{[HA]} \qquad \text{or} \qquad K_b = \frac{[B^+][OH^-]}{[BOH]}$$

Strong acids and bases have K_a and K_b values greater than one because they are completely dissociated. Weak acids and bases have values less than one because the concentration of the acid or base that remains together is greater than the concentration of the separated ions.

pH

In addition to the K_a or K_b to indicate the strength of an acid or base, we often use the pH of the solution to describe whether a solution is acidic or basic. Solutions with a pH less than 7 are considered acidic, and solutions with a pH greater than 7 are considered basic. Solutions with a pH of 7 are neutral. Other relationships between acid and base concentration and pH are

$$pH = -\log [H^+] \quad \text{and} \quad [H^+] = 10^{-pH}$$

$$pOH = -\log [OH^-] \quad \text{and} \quad [OH^-] = 10^{-pOH}$$

$$pH + pOH = 14 \quad \text{and} \quad [H^+] \times [OH^-] = K_w = 1 \times 10^{-14}$$

The determination of the pH and concentration of a weak acid or base solution is complicated by the incomplete dissociation and will be discussed further in the second term of this course. For now, we will worry only about calculating the concentration and pH of solutions containing strong acids and bases.

The pH of a strong acid solution is equal to the negative logarithm (–log) of the hydrogen ion concentration ($pH = -\log[H^+]$) of that solution. Because the strong acid completely dissociates into ions, the acid concentration equals the hydrogen ion concentration.

For example: In a 2.5 M HCl solution, $[H^+] = 2.5$ M and $pH = -\log(2.5) = -0.40$

The same relationship is used to determine the pOH of a strong base. The hydroxide ion concentration equals the concentration of the strong base:

For example: In a 2.5 M NaOH solution, $[OH^-] = 2.5$ M and $pOH = -\log(2.5) = -0.40$. Note that since $pH + pOH = 14$, the pH of this basic solution is 14.4.

A couple of things should be noted about the examples above. First, students often think that pH values cannot be above 14 or below 0. As you can see from the examples given this is simply not true. When dealing with strong acids and bases, the pH values are often very large (for bases) or even negative (for acids). Second, be careful in calculations using the concentration of bases. The negative log of the base concentration gives you pOH, not pH; students often forget this very important second calculation.

The Reaction of a Strong Acid with a Strong Base

In this experiment, the strong acid, HCl, and the strong base, NaOH, are used to create serial dilutions that are combined to form salt solutions of NaCl according to the reaction below:

$$HCl(aq) + NaOH(aq) \rightarrow NaCl(aq) + H_2O(l)$$

These products (a salt and water) are typical of the strong acid–base reaction and can be used to predict the products of other strong acid and base reactions.

For example: $\quad H_2SO_4(aq) + 2KOH(aq) \rightarrow K_2SO_4(aq) + 2H_2O(l)$

Although the reactants have changed, the basic products remain the same: a salt and water.

Prediction of the amount of salt produced is based on the stoichiometry of the reaction and the amount of acid and base provided. In the case of HCl and NaOH, the reaction is one-to-one; therefore, the reactant present in the smaller number of moles limits the number of product moles that can be formed. This number of moles is the *theoretical yield* of the reaction. The *actual yield* will be determined experimentally.

Serial Dilutions

Serial dilution is an important technique in experimentation. The process allows the researcher to create controlled, systematic dilutions from a stock solution. Each successive solution is diluted so that the concentration of solute is changed by a desired ratio. For example, in 1/10 serial dilutions starting with a stock solution of 5 M solute concentration, the serial dilution would be comprised of 5 M, 0.5 M, 0.05 M, 0.005 M… Each solution is reduced to one-tenth of the previous solution's concentration.

For another example, 1/25 serial dilutions starting from a stock solution of 12 M solute concentration would be comprised of 12 M, 0.48 M, 0.0192 M, 0.000768 M…and so on. The greater the ratio of solute concentration change, the greater the difference in concentration for each successive serial dilution. Serial dilutions also conserve supplies because they use the previously made solution as the "stock" solution for the next more dilute solution.

For example:

Using a 0.5 M stock solution, create four solutions of 1/25 serial dilutions. Calculate the resulting concentration for each diluted solution using the following equation:

$$M_1V_1 = M_2V_2$$

where M_1 is the concentration of solution one, V_1 is the volume of solution one, M_2 is the concentration of solution two, and V_2 is the volume of solution two.

$$(1.0 \text{ mL})(0.5 \text{ M}) = (25 \text{ mL})(X)$$

$$X = 0.02 \text{ M}$$

The results indicate that 0.02 M is the concentration of the new diluted solution.

This process is repeated to make the next 1/25 diluted solution. But the stock solution used this time is the newly made 0.02 M solution. Repeating the calculations above:

$$(1.0 \text{ mL})(0.02 \text{ M}) = (25 \text{ mL})(X)$$

$$X = 8 \times 10^{-4} \text{ M}$$

The process continues using each newly made solution as the stock for the next lower concentration until all of the needed dilutions are produced.

As you can see, making 1/25 dilutions very rapidly drops the H^+ concentration to very small amounts. At some point, the concentration of the acid will actually fall below the concentration of H^+ found in pure water. (The pH of pure water is 7, so the $[H^+] = 1.0 \times 10^{-7}$ M). Once you get to this point, you must take into consideration not only the ions from your dilute acid but those from the water dissociation as well.

From the charge balance relationship $[H^+] = [Cl^-] + [OH^-]$, you can substitute for $[OH^-]$ from the water ionization constant:

$$[H^+] = [Cl^-] + K_w/[H^+],$$

and upon rearrangement you get the quadratic equation:

$$[H^+]^2 - [Cl^-][H^+] - K_w = 0$$

This equation can be solved for $[H^+]$ using the quadratic formula.

Procedure

SAFETY NOTES: Acids and bases can cause burns if they come in contact with your skin. Wash your hands immediately if you feel a burning or itching sensation. Use all necessary precautions when handling the acid and base containers.

GENERAL INSTRUCTIONS: This experiment is to be done in pairs; one partner will do the acid dilutions and the other the base dilutions. Be sure to get all of the information from your partner before leaving the lab.

Part I: Preparing Salt Solution

Obtain approximately 15 mL of 0.5 M HCl (red solution) in a clean beaker and 15 mL 0.5 M NaOH (blue solution) in another clean beaker.

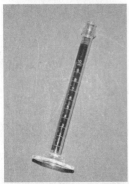

Using a graduated cylinder, measure out and combine exactly 10 mL of NaOH with 10 mL of HCl in a clean 250 mL beaker.

Place the beaker with the acid–base mixture on a hot-plate and set to 100 °C. Allow the mixture to boil gently until all liquid has evaporated.

Place a watch glass over the mouth of the beaker if the salt starts to "spit" out.

While the solution is boiling off, go on to Part II.

Part II: Acid-Base Reactions and pH

Calibrate the pH meter using the three-way calibration procedure provided by your instructor or teaching/lab assistant.

Collect 100 mL of DI water and measure the pH. Record the pH in your lab notebook.

Collect a 25 mL volumetric flask from the front counter.

Using a graduated 1.0 mL pipette, the 25 mL volumetric flask, and either the acid or base, perform four serial dilutions using 0.5 mL of stock solution at each step.

For example: Dilution one, pipette, using the three-way bulb, 0.5 mL of 0.5 M HCl OR 0.5 mL of 0.5 M NaOH into the 25 mL volumetric flask and dilute to the 25 mL mark with DI water. Pour the solution into a clean beaker and then repeat until all dilutions are made.

Calculate the resulting NaCl, H^+, and OH^- concentrations as you did in your prelab work.

Record the pH of the four HCl dilutions and then of the four NaOH dilutions.

Combine the solutions of acid and base by adding the 0.5 M solution of NaOH to the 0.5 M solution of HCl. Record the resulting pH. Repeat the combination of solutions for the other four dilutions. Record the pH for each combination.

Part III: Experimental Yield

Carefully remove the breaker and watch glass from the hotplate using the oven gloves provided. Allow the beaker to cool to room temperature.

Separately weigh the beaker and watch glass containing the salt residue to the nearest 0.001 g and record the values in your lab notebook.

After weighing, clean and dry the beaker and watch glass thoroughly. Reweigh the now empty beaker and watch glass to the nearest 0.001 g and record the values in your lab notebook.

Report Contents and Questions

The purpose of this lab should consist of several well-constructed sentences stating what the experiment was designed to accomplish. Make sure to include statements about concepts and techniques explored in the experiment. The procedure section should reference the lab manual and include any changes made to the lab manual procedure during the lab.

The data section should include three tables. The first table should be for the acid and include the following information: **(a)** concentration of acid after each dilution, **(b)** experimental pH of each dilution, **(c)** theoretical pH of each dilution, and **(d)** percent error in the pH value. The second table should be for the base and include the following information: **(a)** concentration of base after each dilution, **(b)** experimental pH of each dilution, **(c)** theoretical pH of each dilution, and **(d)** percent error in the pH value. The third table is for the combined acid and base solutions and should include: **(a)** concentration of acid and base for each combination, **(b)** experimentally determined pH, **(c)** theoretical pH (hint: think DI water), and **(d)** percent error in the pH value. Also make sure to include the pH of the DI water used for the dilution. Finally, state the balanced chemical equation for this experiment, the theoretical and experimental yield of the salt obtained from heating the acid–base mixture, and percent yield of the reaction.

The experimental pH is determined in lab, but the theoretical pH can be determined for the acid by using $pH = -\log [H^+]$, the $[H^+]$ is the hydrogen ion concentration, which is equal to the concentration of acid you calculated in Part II. For the pH of the base, you must first determine the pOH, using $pOH = -\log [OH^-]$, where $[OH^-]$ is the concentration of the base you determined in Part a. Once you have determined the pOH, you need to subtract it from 14 to obtain the pH. ($pH = 14 - pOH$). The experimental error can be determined with the following formula:

$$\% \text{ error} = \frac{|\text{theoretical value} - \text{experimental value}|}{\text{theoretical value}} \times 100\%$$

Use the balanced chemical equation to determine the theoretical yield of the salt produced. To do this you will need to determine the moles of acid or base (pick one) used in the mixture you heated based on the molarity of the solution and the volume of the acid or base you used (M= moles/L). Use your balanced chemical equation, the reaction's stoichiometric ratios, and the moles you just determined to calculate the moles of salt produced. Finally convert the moles to grams and use the percent yield formula below to determine the percent yield.

$$\% \text{ yield} = \frac{\text{experimental yield (g)}}{\text{theoretical yield (g)}} \times 100\%$$

In the calculation section, be sure to include sample calculations for the following: **(a)** concentration of dilution, **(b)** pH, **(c)** pH from pOH, **(d)** experimental error, **(e)** theoretical yield of salt, and **(f)** percent yield of salt. Also include Excel graphs of the pH versus concentration for both the acid and the base. Remember all graphing guidelines apply here.

The conclusion section should be several paragraphs in length and include the following information: a discussion on the experimental and theoretical pH as well as the percent error for both the acid and the base, with detailed comments on any discrepancies in the values. Within this discussion, explain the trends seen in the graphs you made. Include a discussion on the pH of the combined solutions and the error in the values. Be sure to include a statement about what is expected from these mixtures and why that was not achieved. Finally, discuss the percent yield of salt and any errors that may have caused a yield less than or greater than 100%.

Answer the following questions:

1) Explain how this experiment is related to limiting and excess reagent problems.

2) The acid used in this experiment is monoprotic. What does this mean, and how would the experimental results differ if a diprotic acid had been used instead?

Experiment 9
Laboratory Preparation

Name: _____ **Date:** _____

Instructor: _____ **Sec. #:** _____

Show all work for full credit.

1) Read the background, procedure, and report sections of the lab experiment carefully and develop a hypothesis of what information you expect to gain from the completion of the lab experiment.

2) Create any and all tables you might need to collect data during the experiment and then transfer those tables into your lab manual for use during the lab.

Experiment 9
Prelaboratory Assignment

Name: _____ **Date:** _____

Instructor: _____ **Sec. #:** _____

Show all work for full credit.

1) Constant boiling HCl has a concentration of 11.6 M. Your laboratory assistant is going to dilute this acid to create a stock solution of HCl for you to use in this experiment. What volume of the 11.6 M HCl must the assistant add to water in one-liter volumetric flask so that the concentration of the stock solution will be 0.62 M after dilution to the mark with water?

What is the pH of this stock solution? (Remember the significant figure rules for logarithms!)

You will now use this 0.62 M HCl solution to carry out a series of dilutions, measuring the pH of each diluted sample. Calculate the [Cl⁻] and [H⁺] concentrations and the pH of each sample. (Remember the significant figure rules for logarithms!)

Sample A will be produced by pipetting 1.00 mL of the stock solution into a 25 mL volumetric flask, and diluting to volume with distilled water.

$[Cl^-]=$
$[H^+]=$
$pH=$

Sample B will be produced by pipetting 1.00 mL of sample A into a 25 mL volumetric flask, and diluting to volume with distilled water.

$[Cl^-]=$
$[H^+]=$
$pH=$

Sample C will be produced by pipetting 1.00 mL of sample B into a 25 mL volumetric flask, and diluting to volume with distilled water.

$[Cl^-]=$
$[H^+]=$
$pH=$

Sample D will be produced by pipetting 1.00 mL of sample C into a 25 mL volumetric flask, and diluting to volume with distilled water.

$[Cl^-]=$
$[H^+]=$
$pH=$

Sample E will be produced by pipetting 1.00 mL of sample D into a 25 mL volumetric flask, and diluting to volume with distilled water.

$[Cl^-] =$

$[H^+] =$

pH =

Sample F will be produced by pipetting 1.00 mL of sample E into a 25 mL volumetric flask, and diluting to volume with distilled water.

$[Cl^-] =$

$[H^+] =$

pH =

2) Your laboratory assistant is going to prepare a stock solution of sodium hydroxide (NaOH) for you. What mass of NaOH must the assistant add to water in a one-liter volumetric flask so that the stock solution has a concentration of 0.35 M after dilution to the mark with water?

What is the pH of this stock solution? (Remember the significant figure rules for logarithms!)

You will now use this 0.35 M NaOH solution to carry out a series of dilutions in the same manner as in problem 1, measuring the pH of each diluted sample. Calculate the $[Na^+]$ and $[OH^-]$ concentrations and the pH of each sample A–F.

3) Write the balanced molecular equation for the neutralization reaction between $Ba(OH)_2$ (a strong base) and HCl (a strong acid).

The lab assistant has prepared a 0.378 M solution of $Ba(OH)_2$ and a 0.399 M solution of HCl. What volume of HCl will be required to exactly neutralize 14.0 mL of the $Ba(OH)_2$ solution?

You neutralize 14.0 mL of the $Ba(OH)_2$ solution with this required volume of the HCl solution in a 250 mL beaker, which has a mass of 103.3855 g. You heat the solution to drive off all the water, leaving only solid $BaCl_2$. What mass of $BaCl_2$ do you expect to find in the residue?

You find that the beaker plus the residue has a mass of 104.3628 g.
What is your percent yield of $BaCl_2$?

Experiment 10
The Copper Cycle

Introduction

In this experiment, you will observe a series of reactions that convert a piece of copper metal, via several different copper-containing compounds, back into its original elemental form. Along the way, you will see some striking color changes and gain skills in both recording experimental observations and interpreting those observations in terms of relevant chemical equations. You will also use a simple classification scheme for grouping chemical reactions by type. Lastly, you will practice quantitative lab techniques by trying to recover your original copper sample with minimal loss.

Background

Physical and Chemical Principles

One of the more fascinating aspects of chemistry to a beginning student is the variety of sights, odors, and textures encountered in the laboratory. This experiment demonstrates this aspect by having you take a piece of copper through a sequence of striking changes in color, appearance, and bulk. This experiment will also help develop your lab technique by challenging you to shepherd your copper sample through several chemical changes with minimal loss. Figure 1 shows the cycle of chemical reactions that you will use in this experiment.

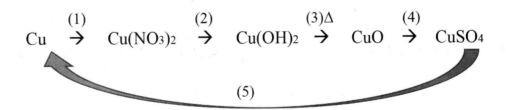

$$Cu \xrightarrow{(1)} Cu(NO_3)_2 \xrightarrow{(2)} Cu(OH)_2 \xrightarrow{(3)\Delta} CuO \xrightarrow{(4)} CuSO_4$$
$$(5)$$

Figure 1: Schematic showing the various intermediate products in the copper cycle.

(1) Nitric acid dissolves nearly all metals to produce nitrate salts. Nitric acid is a powerful oxidizer, much more so than other strong acids like HCl or HBr. In this reaction, the products are copper(II) nitrate, $Cu(NO_3)_2$, water, and nitrogen dioxide (NO_2).

(2) Many metals form insoluble hydroxides, which can be precipitated from solutions of their salts by soluble hydroxides like NaOH or KOH. In this reaction, sodium hydroxide is used to precipitate copper(II) hydroxide, $Cu(OH)_2$, leaving an aqueous solution of sodium nitrate.

(3) Transition metal hydroxides can lose water upon heating and change to oxides. For copper, this reaction happens at an unusually low temperature and is accompanied by a striking color change from blue to black. As copper(II) oxide, CuO, is formed, water is eliminated. This explains the common name of this type of reaction—dehydration.

(4) Metal oxides are basic anhydrides (i.e., a base that has lost one H_2O) and thus form salts when reacted with acids. The salt formed in this step is copper(II) sulfate, $CuSO_4$. The protons from sulfuric acid, H_2SO_4, are used to protonate the oxide and make a water molecule as one of the products.

(5) Chemically active metals, such as zinc, readily displace less active metals (like copper) from their salts. In reaction 5, solid copper is produced, and zinc sulfate is the soluble by-product.

Zinc is also sufficiently active to reduce the protons in an acidic solution to elemental hydrogen.

$$Zn(s) + 2H^+(aq) \rightarrow Zn^{2+}(aq) + H_2(g)$$

This reaction is not explicitly shown in the cycle, but it actually occurs twice. It occurs first with H^+ from excess H_2SO_4 as a side reaction during reduction of $CuSO_4$ and later with HCl, as an intentional step for removal of excess Zn remaining after reduction of Cu^{2+} to Cu. Copper is not sufficiently active to reduce H^+ to H_2. Therefore, we can use the H^+ from HCl to remove zinc (an impurity in our final product) from our recovered copper sample.

Each equation describes one of the colorful transformations you will see, but it is easy to lose sight of their physical meaning if you merely look upon them as a bunch of obscure symbols written on paper. You will soon be on friendlier terms with chemical equations if you try to do three things:

1. Find out what the compounds represented by the formulas actually look like and what their properties are.

2. Find out what class the compounds belong to (e.g., acids, bases, salts) and what the generic properties of each class are (i.e., what type of reactions they participate in).

3. Try to classify the equations and the chemical reactions they represent. Your textbook will have some information on this topic in addition to the information on the following pages.

Chemical Reaction Classifications

Precipitation

Precipitation reactions are those in which two soluble substances are mixed together to form an insoluble compound. In most cases, you will find that the anions in the reaction simply trade places.

$$CaCl_2(aq) + Na_2CO_3(aq) \rightarrow CaCO_3(s) + 2\ NaCl(aq)$$

Acid–Base Reactions: Neutralization

Neutralization covers all reactions of the form acid + base = salt + water. Anhydrides are also included in this type of reaction. The following reactions can all be classified as neutralization reactions.

$$Mg(OH)_2(aq) + H_2SO_4(aq) \rightarrow MgSO_4(aq) + 2\ H_2O(l)$$

$$MgO(s) + H_2SO_4(aq) \rightarrow MgSO_4(aq) + H_2O(l)$$

$$MgO(s) + SO_3(aq) \rightarrow MgSO_4(aq)$$

$$Mg(OH)_2(aq) + SO_3(aq) \rightarrow MgSO_4(aq) + H_2O(l)$$

Hydration

The hydration of acid or base anhydrides will lead to the formation of acids or bases. The reverse, dehydration of acids or bases to form anhydrides, is also true.

$$CaO(s) + H_2O(l) \rightarrow Ca(OH)_2(aq)$$

$$H_2CO_3(aq) \rightarrow CO_2(g) + H_2O(l)$$

Displacement

This type of reaction occurs when an acid (or base) is displaced by another acid (or base).

$$H_2SO_4(aq) + 2\ NaCl(aq) \rightarrow Na_2SO_4(aq) + 2\ HCl(aq)$$

Oxidation–Reduction Reactions

This class is easy to recognize because it, unlike the other types, involves a change in oxidation number. It includes all reactions where an element is changed to a compound, or vice versa.

$$Mg(s) + O_2(g) \rightarrow MgO(s)$$

The Experiment

Notebook

Students are expected to arrive at lab prepared to start the experiment. The lab notebook should contain preliminary notes sufficient for any person to complete the experiment without the lab manual. Instructors and/or TAs will be checking notebooks to see that sufficient detail is present to complete the lab.

Miscellaneous Notes

All equipment should be clean before starting your experiment. This lab has several steps that require rinsing with hot deionized water. It is a good idea to keep a large beaker of water on a hot plate. Adjust the hot plate to keep the temperature of the water around 70 °C. It may be a good idea to share this resource with a neighbor. All reactions in this experiment are to be carried to completion. Be sure to add an excess of every reagent. The amounts specified in the procedure are in general more than enough. In some cases, side reactions may consume part of the reagent added, and you should check that the reagent you add is indeed present in excess. This is a quantitative experiment, with an expected yield close to 100%. Consequently, try to lose no more than 0.2% of the Cu in each step, as estimated from the volume of the solution or solid you are working with. Also minimize the amount of impurities (e.g., dust, dirt, moisture, residual reagents) in your final sample. Use deionized water in this experiment. Keep your wash bottle filled with deionized water.

Safety Notes

Concentrated nitric acid, HNO_3, is hazardous. Wear gloves while handling it. This acid attacks skin in a matter of seconds and produces painless but unsightly yellow stains. If it is left on your skin for more than 10 min, it may produce painful blisters or burns. Handle nitric acid only in the fume hood. Use the labeled beaker in front of the nitric acid stock bottle to pour the nitric acid into your graduated cylinder. Nitric acid has a low surface tension and tends to run down the sides of containers—be careful when pouring it. Rinse your hands with plenty of water after handling nitric acid. Sodium hydroxide solutions attack skin, eyes, and wool. Wash off spills with plenty of water. Nitrogen dioxide fumes will appear brown. They are toxic and must be removed in the hood. Acetone is flammable and its vapor, although not toxic, is not healthy. Use it in the hood and keep it away from open flames.

Procedure

1. Dissolution of copper

Using a Kimwipe, pick up a precut, precleaned piece of Cu wire from the dispenser. Mass the wire on an analytical balance to the nearest 0.0001 g. If the balance reading fluctuates, check Appendix C or ask your laboratory instructor for help. After recording its mass, place the wire in a clean, dry 250 mL beaker.

Measure out about 4.0 mL of concentrated HNO_3 (16 M). In the fume hood, add the nitric acid to the beaker containing the wire. Make sure the wire is completely submerged in the acid solution. Cover the beaker with a watch glass, convex side down. A lively reaction should occur. Occasionally, the reaction can be somewhat sluggish. Check with your laboratory instructor to see if you need to heat the beaker gently over a very small flame in the fume hood to move the reaction along. Record your observations. Swirl the solution around while holding onto the watch glass until the Cu has completely dissolved. What is in the solution now that the reaction is complete?

You should see a harmful brown gas inside the beaker. What is it? In the next step, you will remove the watch glass to recover any sample splashed on the watch glass. Be careful and try not to inhale the gas! The fume hood will remove the gas through a vent away from you.

Next, use the following technique to try to recover any sample sprayed onto the watch glass. Hold the watch glass by the edges, being careful not to get any HNO_3 on your hands. With your wash bottle, rinse the underside of the watch glass, letting the rinse run into the beaker.

Similarly, rinse the walls of the beaker. Then add distilled water until the beaker is about half full. Return to your bench and continue with the next step of the experiment.

2. Precipitation of copper(II) hydroxide

In a 50 mL graduated cylinder, measure out about 30 mL of 3.0 M NaOH. While stirring the dissolved copper(II) solution in the 250 mL beaker with a glass rod, slowly add NaOH to precipitate $Cu(OH)_2$. What else is in the solution?

Let the precipitate settle and then record your observations and note how much bulkier the precipitate is than the initial Cu sample. The extra atoms (two O and two H) in $Cu(OH)_2$ do not account for the bulk, as they are only slightly larger than the copper. Most of the bulk comes from water trapped in this gelatinous precipitate.

3. Conversion to copper(II) oxide

Place the beaker containing the precipitate on a wire gauze atop a ring that has been clamped to a ring stand. Using a Bunsen burner, gently heat (with a proper cone-shaped flame) the solution in the 250 mL beaker just to the point of boiling while stirring gently to prevent "bumping" (i.e., an explosive formation of a large steam bubble in a locally overheated area). A bump can cause threefold damage: a second-degree burn from hot liquid, an alkali burn from NaOH, and loss of your precious sample. (*The lab instructor may decide to use a hotplate to heat the reaction rather than a Bunsen burner, so please consult with him/her prior to starting this step!!*)

Record your observations and interpret the color change. Continue heating for a few minutes after the transformation is complete to coarsen the precipitate. The solid is the precipitate that is formed from the heating process. It creates larger solid pieces and allows it to settle quicker because the chunks are now heavier. Remove the burner, continue stirring for a minute or so, then set the beaker aside.

Start heating about 200 mL of distilled water in a 400 mL beaker. This will be used to wash the precipitate. Let the CuO settle for about 10 min or so until it is no more than 1/4 the volume of the liquid. Decant the supernatant liquid into a clean 400 mL beaker, being careful not to lose more than a trace of CuO. Put a piece of white paper under the beaker to help you see if any black particles are being carried over.

Remember, you can afford to lose up to 0.2% of your sample during this experiment. If you lose a large amount of sample while decanting the solution, let it settle in the second beaker, decant again, and combine the recovered solid with the main batch. Do not worry about a few persistent floaters. Such floaters might appear if you started with a dirty wire or forgot to handle the wire with the Kimwipe tissue.

Add 150 mL of the wash water you prepared earlier, let the CuO settle, and decant once more. What is removed by this decanting and washing procedure? Check with red litmus paper and see whether your answer is consistent with the litmus test.

4. Dissolution of cupric oxide

Add 15 mL of 6.0 M H_2SO_4 while stirring. Record your observations. What is in the solution now?

5. Reduction of Cu^{2+} to Cu

Mass out approximately 3 g of zinc. Initially add 1 g of zinc to your reaction mixture. Add the remaining portions in 0.2 g increments, as needed, one at a time, to the solution in the beaker. The 0.2 g portions should be added when the zinc from the previous addition is almost completely gone. Cover the beaker with a watch glass to catch acid spray and swirl occasionally to bring fresh solution into contact with the zinc. It is possible that only 2.5 g of zinc will be needed.

Part of the Zn reduces Cu^{2+} to Cu while the rest is destroyed by a side reaction with H^+, yielding H_2 gas. The formation of H_2 is actually beneficial, since it prevents the buildup of a solid Cu film that would coat the Zn and stop the reaction. Bubbling is a good sign at this point. It shows that there is sufficient Zn present to reduce any Cu^{2+} and sufficient acid to prevent the Cu from "gumming up" the Zn.

Wait for about 10 min. During this time, look at the bulleted sections below and follow the instructions that correspond to your conditions.

- If the blue color still persists after 10 min or after the bubbling has stopped, then the reaction—$Cu^{2+}(aq) \rightarrow Cu(s)$—may have stopped before all the $Cu^{2+}(aq)$ was converted to $Cu(s)$. This is why the solution is still blue.

This situation can be caused by the exhaustion of either the acid or zinc. Perhaps you did not remove all of the NaOH solution during the decanting and washing procedure. NaOH solution dilutes and partially neutralizes the acid that should have reacted with the Zn. It is also possible that too much acid was present, and zinc was used up, making H_2 instead of copper metal.

To remedy this, first add 0.5 to 1.0 g of Zn to the solution. Up to 2.0 g may be needed if the solution is still quite blue. If the bubbling resumes, then you had simply run out of Zn. If the solution bubbles only slightly or not at all, then you are low on acid. Add another 10 mL of 6 M H_2SO_4 and let the reaction run until the solution is absolutely colorless. Proceed to the next bulleted step below.

- If the blue color completely disappears, the Cu^{2+} was completely reduced before either the Zn or the H_2SO_4 were exhausted. To confirm the color disappearance, put the beaker on a piece of white paper, with a beaker of water alongside as a control.

- If the bubbling still occurs after the color has disappeared, let the bubbling run for at least another 5 min. Even a seemingly colorless solution can still contain trace amounts of Cu^{2+}.

Now, test for completeness of reduction. Decant at least 1/3 of the solution into a clean 250 mL beaker (beaker 2), taking care not to carry over any Cu particles. Put about 1.0 g of Zn into beaker 2 and check after a few minutes whether any more Cu has separated. The copper might be in the form of very fine black rather than red particles, or even of a gray film on the Zn, dulling and darkening its normal silvery color. If any Cu appears, return the entire contents of beaker 2 (solution and Zn) to the original beaker and let the reaction run its course for at least for another 5 min or until a similar test later is negative. Regardless, proceed to the next step in which excess zinc is removed.

Removal of Excess Zinc

To get rid of any excess Zn that remains with your recovered Cu, first decant any liquid. Be careful not to lose any solid (Cu and Zn). Add 10 mL of 6 M HCl to the Cu and Zn. After the bubbling has slowed down, cautiously heat the beaker on the hot plate until the characteristic fine bubbles of H_2 are no longer evolved. It is a good idea to cover this reaction with a watch glass.

Decant the liquid and wash the solid with 100 mL of distilled H_2O. Be careful to rinse the walls of the beaker during the washing. Repeat this washing process a second time. For the third wash, add 100 mL of distilled H_2O and boil gently on the hot plate for 5 min to remove any traces of acid trapped in the Cu. Dip a clean stirring rod in the hot solution and wipe the stirring rod on a piece of blue litmus paper. The solution should no longer be acidic. Besides the acid, what else are you trying to get rid of by these washes?

Washing with Acetone

Carefully mass an evaporating dish on an analytical balance. Using a glass rod, transfer the wet Cu from the beaker into the evaporating dish. Rinse the remaining Cu out of the beaker into the evaporating dish with a distilled water wash bottle. Drain the water from the dish by careful tilting.

To displace any remaining water, wash the Cu with 5 mL of acetone. Add the acetone to the dish while swirling around so that the acetone penetrates the copper granules and then decant. Repeat the above step with a second 5 mL acetone wash. When decanting, leave no more than a drop or two of the liquid at the bottom of the dish. You may use a pipette if necessary to remove any remaining acetone without removing the Cu. If there is any remaining Cu, wait a few minutes for the acetone to evaporate. Reweight this process and record in your lab notebook.

Waste Disposal

Place all waste materials except acetone in the aqueous waste container; acetone is collected in a separate container marker "Acetone—Organic Waste."

Calculations

This section describes the one calculation needed to successfully complete the laboratory assignment. Read through it carefully and make sure that you understand it before you leave the laboratory.

The percent recovery is a gauge of how well you were able to keep track of the copper you were issued at the outset of the experiment. If every step of the procedure went as planned, you should have recovered all of the original copper, that is, you had 100% recovery. Unfortunately, unexpected things happen along the way and, chances are, your recovery will not be equal to 100%.

The percent recovery is calculated through the following equation:

$$(m_f/m_i) \times 100\%$$

where m_i and m_f are, respectively, the initial and final masses of your copper sample. The final mass of the copper sample is calculated by difference by subtracting the mass of the evaporating dish, m_d, from the mass of the dish with the sample in it, m_{sd}, via:

$$m_f = m_{sd} - m_d$$

Answer the following question:

1) It is common for students to find that their percent recovery is greater than 100% for the copper cycle experiment. Can this cycle really make more copper from the reagents used? Provide two explanations (using complete sentences, of course) for why someone might have a percent recovery greater than 100%.

Experiment 10
Laboratory Preparation

Name: _____ Date: _____

Instructor: _____ Sec. #: _____

Show all work for full credit.

1) Read the background, procedure, and report sections of the lab experiment carefully and develop a hypothesis of what information you expect to gain from the completion of the lab experiment.

2) Create any and all tables you might need to collect data during the experiment and then transfer those tables into your lab manual for use during the lab.

Experiment 10
Prelaboratory Assignment

Name: _____ Date: _____

Instructor: _____ Sec. #: _____

1. Balance the equations given below and classify each as a precipitation, acid–base, hydration (or dehydration), or oxidation–reduction reaction.

   a) _____$NaOH(aq)$ + _____$FeCl_3(aq)$ → _____$Fe(OH)_3(s)$ + _____$NaCl(aq)$

   Reaction type: _____

   b) _____$Al(s)$ + _____$HCl(aq)$ → _____$AlCl_3(aq)$ + _____$H_2(g)$

   Reaction type: _____

   c) _____$KOH(aq)$ + _____$H_3PO_4(aq)$ → _____$H_2O(l)$ + _____$K_3PO_4(aq)$

   Reaction type: _____

2. For the scenarios listed below, state whether you would expect the error to result in a high, low, or unaffected percent yield. Provide a brief explanation for each of your answers.

   a) In the precipitation of copper(II) hydroxide step, insufficient NaOH is added.

   High, low, unaffected: _____

   Explanation: _____

   b) In the dissolution of copper(II) oxide step, excess H2SO4 is added.

   High, low, unaffected: _____

   Explanation: _____

Experiment 11
Atomic Spectra

Introduction

Have you ever wondered how scientists know what elements are present on other planets? Believe it or not, the information regarding the elemental makeup of each of the stars, planets, and other heavenly bodies is being broadcast right to us on a daily basis. Each of these bodies has what is called an *emission spectrum* that can be read from far away using a device called a *spectrometer* (and a really powerful telescope, too.) The spectrometer is able to separate the colors of the light being emitted by the star into discrete lines. The origin of these lines is discussed further in both your textbook and in the background section for this lab. Each element has a unique set of spectral lines; with a sophisticated computer program that can separate the individual spectra from the total array of spectral lines, you can identify the elements present in galaxies far, far away.

In today's lab, we will be working a little closer to home by learning to use spectroscopes and to interpret the meaning of a spectrum. You will experimentally obtain the spectra of several unknowns and then, using known "bright line" spectra for comparison, identify unknown ions.

Background

During the past century, scientists like Sir J.J. Thomson, Ernest Rutherford, and Niels Bohr studied atomic structure. These studies suggested several different models, each attempting to further describe the internal structure of the atom. The Bohr model of the atom is especially easy to visualize. Chapter 5 of your textbook describes the development of these models and also describes the calculation of energy and wavelengths.

The Hydrogen Atom

The structure of the hydrogen atom is very simple: one proton and one electron, a model known for some time before Bohr. It was also known that the electron is negatively charged, and the proton is positively charged. This created some intellectual difficulty for scientists: They did not understand why the electron doesn't spiral into the proton—since negative charges are attracted to positive charges—and completely destroy the atom. But, since hydrogen is a stable atom, this attraction was obviously not the only rule that governed the behavior of the atom.

Learning how atoms were able to maintain this stability was a very important problem. The most advanced physics of the time, Maxwell's *theory of electromagnetism*, predicted the instability of atoms. Because of this, there was clearly a very fundamental flaw in scientists' understanding of the universe.

Max Planck, Niels Bohr, and Energy

Prior to Niels Bohr's model of the hydrogen atom, Max Planck had postulated that light was composed of photons that carried quanta, discrete packets of energy. Utilizing this notion, Bohr theorized that since there were well-defined states in which atoms could exist, then photons with just the right energy could cause transitions between these states. In fact, the reason that the electron did not spiral into the proton was that the atom was actually in one of these stable states. Therefore, Bohr's theory also stated that the electrons were quantized in energy levels.

Quantization

The idea of quantization is often difficult to accept. In the macroscopic world, things are continuous; they can take on any value. For example, a car can travel 17 mph, 18 mph, or 17.5 mph; it is not restricted to only integer values. The speed is perceived on such a large scale that we cannot tell it is quantized, so it appears to be continuous. However, when examined more closely, this continuity no longer holds. The speed of the car is, in fact, quantized, but since the difference between two adjacent levels is so small, quantization cannot be observed.

Quantization actually means that only specific values are allowed. For example, compare a set of steps and a ramp. Since potential energy is a function of height, the potential energy is quantized for the steps. An object can rest on only one step or another; it cannot be between steps for a considerable amount of time. Thus, the potential energy increases only in increments of the height of the step. On the other hand, the ramp is not bound by such restrictions. For the ramp, the potential energy is continuous.

The Predictions and Mathematics

To fully understand Bohr's predictions, some math is involved. Bohr's model predicted that the energy of a hydrogen atom was quantized and that the energy of the atom was dependent on the principal quantum number (n), which is always an integer. The exact dependence of the energy on this value is complicated to derive, but it turns out that the energy, often abbreviated E_n, is given by the equation shown below, where β is a constant based on Planck's constant, the mass of the electron, and the charge of the electron. The value of β is ~2.18×10^{-18} Joules.

$$E_n = -\beta\left(\frac{1}{n^2}\right)$$

Note that the energy is negative. If E_n is thought of as a measure of the energy of attraction between the electron and proton, we can see that the larger the value of n, the less negative the value of E_n. This means that there is less attraction between the electron and the proton, so the distance between them increases. When n is very large, the value of E_n approaches zero, and there is no attraction between the electron and the proton. Therefore, the electron is free to go wherever it wants—it is no longer bound to the proton.

It should also be noted that the smallest value of n is one; thus, the lowest energy state for hydrogen atoms is the value of β, not zero as one might expect. In fact, the only time the energy of the electron is zero is when it approaches an infinite distance from the nucleus.

Rydberg Equation

By itself, the equation given above is not very useful. Because there is no energy parameter used to measure the absolute energy, only energy differences can be measured. Consider two possible states of the hydrogen atom that have quantum numbers n_I and n_F. (Note: n_I and n_F stand for n initial and n final, respectively. These are the initial energy level from which the electron transfers and the final energy level where it ends up.) If we consider a transition from the n_I level to the n_F level, the change in energy (ΔE_{level}) will be:

$$\Delta E_{level} = E_{final} - E_{initial} = \beta\left(\frac{1}{n_F^2} - \frac{1}{n_I^2}\right)$$

The change in energy values calculated using this equation is positive if absorption is occurring (i.e., when the electron moves from a lower energy state to a higher energy state) and negative if emission is occurring (i.e., when the electron is moving from a higher energy state to a lower energy state). This energy is equal to the energy of the photon that was either absorbed or released by the atom. The energy of a photon with a frequency ν is simply the value of that frequency multiplied by Planck's constant (h), which is 6.626×10^{-34} J-sec.

$$E = h\nu$$

Furthermore, the frequency is equal to the speed of light (c), 3.00×10^8 m/sec, divided by the wavelength (λ) of the photon in meters:

$$\nu = \frac{c}{\lambda}$$

Combining these relationships, we can solve for the specific wavelength of light that is associated with a given transition. This is known as the *Balmer-Rydberg equation* and is shown below.

$$\frac{1}{\lambda} = -R\left(\frac{1}{n_F^2} - \frac{1}{n_I^2}\right)$$

In this equation, R is known as the Rydberg Constant (1.097×10^{-2} nm^{-1}), takes into account Planck's constant as well as the speed of light, and has units that are more practical for calculating wavelengths. Although calculations may result in negative values of wavelength, only positive values are reported.

This Week's Adventure

The Rydberg equation predicts a measurable value, the wavelength of light emitted by a transition between two states. For this experiment, n_F will always be 2, the value corresponding to the visible spectrum. Therefore, by substituting values for n_I that correspond to the visible spectrum of hydrogen, we can predict the wavelengths we should observe.

In this experiment, we will excite helium and hydrogen atoms by subjecting them to an electrical potential in a lamp. We will also excite a variety of cations by heating them in a flame. Each of these lamps or flames will be marked as unknowns. It will be your job to use your knowledge of their known wavelengths to identify them.

The Spectroscope

The spectroscope used in the experiment has a diffraction grating as its central component. A diffraction grating "bends" light much like a prism. The angle at which the light is bent depends upon the wavelength of light entering the grating.

Before we can start accurately measuring an unknown spectrum, we must first calibrate the spectroscope. The known spectrum of helium that will be used is given in the table below, to calibrate your spectroscope. Once plotted, the measured values of the helium lines against the known values, we can use that graph to determine the actual wavelengths of the unknown lamps and flames provided.

Color	Wavelength (nm)
Violet	388.9
Blue	468.6
Green	501.6
Yellow	587.6
Orange	667.8
Red	706.5

Procedure

SAFETY NOTES: If the flame is somehow extinguished while doing the flame spectra, turn off the gas and call the instructor immediately.

Be careful not to get too close to any of the setups. The lamps have enough voltage to give a very serious shock. There is no reason to touch any of the setups.

Some of the lamps also emit ultraviolet light, which can damage your eyes. Do not look into a lamp for a long period of time.

GENERAL INSTRUCTIONS: Students work in pairs for this experiment. *Since you are calibrating the instrument, the same spectroscope must be used in each part.*

Part I: Spectroscope Calibration

Obtain a spectroscope.

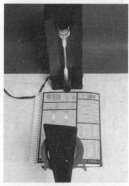

Obtain a ring stand. Set the ring stand up in front of the helium lamp, using your lab notebook to space the ring stand from the lamp (i.e., the legs of the ring stand should be exactly one notebook length away from the lamp).

Set your spectroscope up on the ring stand. Focus the spectroscope so that light from the helium lamp comes in directly through the vertical slit. Look through the eyepiece where the grating is located and find the bright visible lines on the scale to the right of the slit. Read the scale accurately. You should be able to read the numbers and marks, as well as estimate between the marks, so each number you record should have three significant figures. Make sure that you do not move the spectroscope once you have aligned it with the light source. Have your partner record your readings. Now let your partner read the spectroscope while you record. He/she should read the spectroscope in

the exact same alignment that you did. Check to see if your readings agree. You may decide that several attempts are necessary to get the most reliable readings.

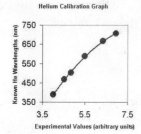

Using graph paper from the back of your lab notebook, create a graph of the known wavelengths of helium versus the wavelengths of the same colors you just observed experimentally. Make a best fit line to your data.

Part II: Identifying the Unknowns

Use the spectroscope that you just calibrated.

Set up your spectroscope in front of one of the unknown light sources. Record all the colors and wavelengths of each line you observe.

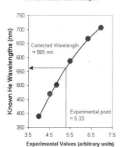

On your calibration graph, on the *x*-axis, find the value of the first of your observed lines. Now, using a vertically aligned straight line, find the point on your best-fit line that corresponds to that value. Now find the point on the *y*-axis that also corresponds to that point on the line and record that value in your notebook.

Repeat the steps above for the observed lines of all of the unknowns.

Report Contents and Questions

The purpose should be in paragraph form describing what your experiment was designed to accomplish and the criteria used to determine success. These paragraphs should include both concepts and techniques. The procedure section should reference the lab manual and note any changes that were made in the procedure during experimentation.

The data section for this report should have two data tables. The first table should have your helium calibration data. This should include **(a)** the known He wavelengths and **(b)** the experimental wavelengths you observed. The second data table is for your unknown data. This should include **(a)** the unknown light source, **(b)** the observed values from the unknown light source, **(c)** the corresponding colors, **(d)** the corrected wavelength, and **(e)** the identity of your unknowns (Compare your corrected wavelengths to those calculated in your prelab assignment).

The helium calibration, unknown light source, observed values, and corresponding colors are all data that can just be copied from your lab notebook. The corrected wavelength for the unknown can be determined from the helium calibration graph.

The calculation section should include the calibration graph for the helium data. This graph should have the experimental value for helium on the *x*-axis and the known helium wavelength on the *y*-axis. All graphing guidelines apply.

The conclusion section should be several well-developed paragraphs that include the identity of each unknown light source, a detailed explanation of how the identity of each light source was determined, a discussion of possible errors, and any changes you would make to improve the experiment should you have the opportunity to repeat it.

Answer the following questions:

1) What is the difference between an emission and an absorption spectrum?

2) Each element has a unique set of wavelengths associated with it. How are these different wavelengths generated?

3) How does NASA use spectroscopy to explore far-away planets?

4) We say that the Sun is a yellow star. What does this really mean?

Experiment 11
Laboratory Preparation

Name: _____ Date: _____

Instructor: _____ Sec. #: _____

Show all work for full credit.

1) Read the background, procedure, and report sections of the lab experiment carefully and develop a hypothesis of what information you expect to gain from the completion of the lab experiment.

2) Create any and all tables you might need to collect data during the experiment and then transfer those tables into your lab manual for use during the lab.

Experiment 11
Prelaboratory Assignment

Name: _____ Date: _____

Instructor: _____ Sec. #: _____

Show all work for full credit.

1) You have just carried out the calibration of your spectroscope's diffraction grating, and you get the values shown in the table for the visible emission lines of helium. Plotting these values gives the graph shown below.

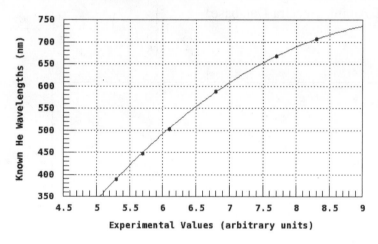

Color	Wavelength, nm	Spectrometer Reading
Violet	388.9	5.3
Blue	447.1	5.7
Green	501.6	6.1
Yellow	587.6	6.8
Orange	667.8	7.7
Red	706.5	8.3

What would be the wavelength of spectral lines that give the following readings on this spectrometer?

Spectrometer reading: 6.5; wavelength:
Spectrometer reading: 7.5; wavelength:
Spectrometer reading: 8.1; wavelength:

What would be the frequency of these spectral lines?

Spectrometer reading: 6.5; frequency:
Spectrometer reading: 7.5; frequency:
Spectrometer reading: 8.1; frequency:

What would be the energy of a photon from each line?

Spectrometer reading: 6.5; energy of one photon:
Spectrometer reading: 7.5; energy of one photon:
Spectrometer reading: 8.1; energy of one photon:

What would be the energy of a mole of photons from each line?

Spectrometer reading: 6.5; energy of one mole of photons:
Spectrometer reading: 7.5; energy of one mole of photons:
Spectrometer reading: 8.1; energy of one mole of photons:

2) The following atomic line spectra display six of the more prominent lines observed in the visible region for the indicated element. The number by each line represents its relative intensity. A photograph of the actual spectrum of the element may show many more lines, or sometimes fewer lines, depending on the overall intensity of the spectrum. If two lines are within 1 nm of each other, they are shown as a single line. You can find a complete listing of the lines for each element at the NIST Atomic Spectra Database.

Some spectral lines of Krypton:

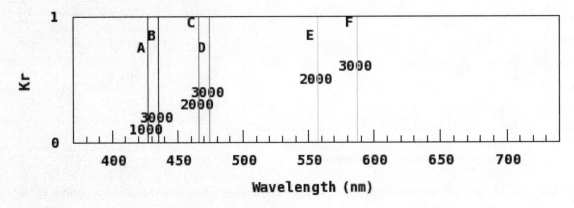

Identify the line in the spectrum of Kr with each of the following characteristics:

Light with a photon energy of 256.8 kJ/mol.
Light with a frequency of 7.014×10^{14} s^{-1}.
Light with a photon energy of 3.566×10^{-19} J/photon.

Experiment 12
Halogen Reactions

Introduction

One of the most basic facets of being a chemist is being able to look at two chemical species and predict whether they will react. You will find that most reactions, even the more elaborate organic reactions in your future, are based on the very simple chemical properties that you are learning about in this class.

One of the properties explored in this experiment by the use of the halides and the halogens is electronegativity, a bonded atom's ability to attract electrons toward it. Other concepts illustrated by the reactions of the halogens are polarity of bonds and of molecules, and solubility. These three concepts are interrelated. The presence of an electronegative atom in a molecule is required for a molecule to be polar. (NOTE: Other factors such as structural symmetry and the magnitude of the atom's electronegativity also play a role and will be discussed in greater detail later, but the presence of that electronegative atom is key.) A molecule's solubility depends on its polarity. Thus, knowledge of the simple concept of electronegativity allows you to make predictions about a molecule's solubility.

Because this lab is qualitative in nature rather than quantitative, there are no calculations and no measurements. Rather, we will be collecting colorimetric evidence; that is, you will be recording the colors observed for each of the halogen solutions placed in the test tubes.

Background

One of the most useful things about studying chemistry is discovering the amount of useful information contained in the periodic table. Many of these periodic properties of the elements are discussed in your textbook.

In today's experiment, some of the properties of the Group 7A (Group 17) elements, known as halogens, and their compounds will be explored. The solubility properties of the halogens will be used to predict their reactions.

The more electronegative an element is, the more it attracts electrons. Group 17 atoms that have become ions by gaining an extra electron, such as F^-, Cl^-, Br^-, and I^-, are called *halides*. Note that the chlorine atom in NaCl, sodium chloride, is a halide—specifically, a chloride. Also note that anions such as halides must always be paired with cations when found in the formula for a binary ionic compound. Group 17 atoms in their natural diatomic state, such as F_2, Cl_2, Br_2, and I_2, are called *halogens*. In this experiment, the relative *electronegativities* of the halogens will be determined.

If a solution containing a halide, X^- (ionic), is added to a solution of a different halogen, Y_2 (non-polar covalent), there are two possibilities. When element Y is more electronegative than element X, Y_2 will take the electron from X^-, leaving X_2 as a *halogen*. On the other hand, when Y_2 is less electronegative than X^-, no reaction will take place, and Y_2 remains as the halogen. In terms of balanced equations:

$$2X^- + Y_2 \rightarrow X_2 + 2Y^-$$

or

$$2X^- + Y_2 \rightarrow \textbf{no reaction}$$

Polar and Non-Polar Solvents

The properties of two solvents, water and hexane, are useful in sorting out what happens in this type of reaction. Water is a polar solvent, and it will solvate *polar* species, whether they are ionic or molecular. (*like dissolves like*) This means that a polar molecule (one that has a dipole moment) or an ionic compound may dissolve in water. A diatomic molecule is polar if the two atoms have different electronegativities.

Non-polar solvents solvate *non-polar* molecules. Hexane (C_6H_{14}) is an organic molecule that is non-polar. Since water is polar and hexane is non-polar, the two do not mix. When combined, two distinct, colorless layers are formed with water, the denser liquid, on the bottom.

If colored substances are added to a test tube containing water and hexane, the polarity of the substances can be determined. If they are non-polar, they will color the hexane layer. If the colored substances are polar, the color is observed in the water layer.

For the first reaction described above, if Y_2 is green before any reaction takes place, the hexane layer is green because non-polar compounds will reside in the hexane layer. As the reaction proceeds, the green disappears from the hexane layer because the Y_2 molecules are reacting and disappearing. The hexane will then take on the color of X_2. In the second case, where X is more electronegative than Y, the more electronegative atom already has the electrons, so no reaction will occur. Since no reaction occurs, the hexane layer will remain green, the color of Y_2.

Whenever a color change occurs, this is a clue that a reaction is taking place. In this experiment, you will observe halogens in aqueous and hexane solution, the solubility of halide ions, and the changes in these observations as you mix halogens with halides.

Procedure

SAFETY POINTS: All halogen solutions are considered somewhat dangerous. Iodine can burn the skin, and chlorine and bromine are corrosive. The volatility of chlorine and bromine increases the danger of inhaling fumes. Avoid all contact with these compounds. If a spill of any of the halogen solutions occurs, call the instructor immediately.

Any solution containing bromine must be used only under the hood and should be disposed of in the proper waste container. The only solutions that may be disposed of in the sink are the water solutions of the halide salts. All other solutions must be discarded in the proper waste container.

Although hexane is not considered a toxic substance, it is an organic solvent, which cannot be disposed of via the sewer. Therefore, all solutions that contain hexane must be disposed of in the organic waste container.

GENERAL INSTRUCTIONS: At least 12 clean test tubes will be needed for today's experiment.

Part I: Solubility Testing of the Halides

This portion of the experiment tests the solubility of the *halides*. Two sets of halide salts are available, sodium salts (Na^+) and potassium salts (K^+). Choose one of the sets. Get together with another student, who will test the other set to share data.
Add a *small* amount (the size of a small pea) of the *solid chloride* salt to a test tube. Note the appearance and other properties in your notebook.

Now add 2 mL of distilled water and agitate the test tube. Record your observations. Repeat for each of the other *solid halide salts* in the set.

Repeat Steps 1 and 2 using hexane as the solvent instead of water. NOTE: Any solution that contains hexane should be disposed of in the waste jar under the hood.

Part II: Solubility Testing of the Halogen

NOTE: If you are not good at describing subtle differences in colors, you may use the crayons provided to make accurate color notes on the two layers of the test tubes for the remaining tests. Simply draw a diagram of the test tube in your notebook and match the color of the experimental layers using the crayons. Remember, sometimes a picture is worth a thousand words.

The solubility of the *halogens* will be tested next. Add 1 mL of the aqueous (water) solution of iodine (I_2) to a test tube. Iodine is a solid at room temperature. Can you hear the solid crystals rattling in the bottom of the dropper bottle?

Do not measure the amount in a graduated cylinder. An approximate amount is good enough, and 20 drops is roughly 1 mL. Be careful not to spill any of the halogen solutions on your skin or clothes, as it will stain. Record the appearance of the aqueous solution.

Now add 1 mL hexane and gently tap the bottom of the test tube to mix. Be sure it mixes; it may take a more vigorous shake, but start gently. Don't spill, but *do not* put your finger over the top and shake. Record all observations of both layers. When your observations are complete, put the solutions into the labeled waste jar in the hood. Rinse the test tube and put the rinsing into the jar, too.

Bromine is a liquid at room temperature and is quite volatile (i.e., it evaporates easily). The aqueous solution of bromine (Br_2) is under the hood. Do not take any test tube containing bromine out of the hood. When finished with the bromine solutions, pour them into the waste jar in the hood and then rinse the test tube with water from a wash bottle.

Add 1 mL of the bromine solution to a test tube. Record the appearance of the aqueous solution. Just as before, add 1 mL of hexane and mix. Record all observations.

Cl_2 is normally a gas at room temperature. Since aqueous solutions of chlorine are not stable for very long (It's the equivalent of a soda going flat due to the loss of CO_2.), each student will make a solution of chlorine (Cl_2).

Add 1 mL hexane to a test tube.

Now, add 0.5 mL Clorox and 0.5 mL 1 M HCl to the test tube. Do not breathe the bubbles that are given off–this is chlorine gas. Note the color of each layer. When finished, pour the mixture into the waste jar and rinse the test tube.

Part III: Electronegativity Testing

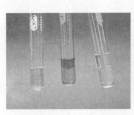

Now that baseline observations of what each of the ions and molecules looks like in water and in hexane, this set of reactions will look at all possible combinations of two reactants. This should allow relative electronegativities of the halogens to be confirmed. Remember that you are now doing a reaction, so you must make sure that your solution layers are thoroughly mixed to make sure the reactants come in contact with each other.

Add 0.5 mL Clorox to 0.5 mL 1 M HCl to make $Cl_2(aq)$ and then add 1 mL hexane and mix. To this, add 1 mL of the water solution of I^-. Mix well and record your observations. Dispose of the solution in the waste jar.

Repeat using the water solution of Br^- instead of the I^-.

Add 1 mL of the $Br_2(aq)$ to 1 mL hexane, mix, and add 1 mL of the water solution of I^-. Mix well and record all observations. Again, dispose of the solution in the waste jar.

Repeat using the water solution of Cl^- instead of the I^-.

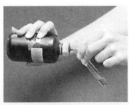

Add 1 mL of the $I_2(aq)$ to 1 mL hexane, mix, and add 1 mL of the water solution of Cl^-. Record all observations. Mix well and record all observations. Again, dispose of the solution in the waste jar.

Repeat for the water solution of Br⁻ instead of the Cl⁻.

You have now added each of the *halide* solutions to each of the *halogen* solutions. By noting the color of the hexane layer in each reaction, you can determine whether a reaction took place, and you can organize an experimental ranking of the electronegativities of the halogens. Before cleaning up your glassware and solutions, ensure that your data make sense. Talk with your instructor about repeating any tests that give contradictory conclusions.

Report Contents and Questions

The purpose should be several well-constructed sentences describing what your experiment was designed to accomplish and the criteria used to determine success. The procedure section should reference the lab manual and note any changes made to the experiment. For this experiment, the procedure section should also state whether you tested potassium salts or sodium salts, and whose data you used for the other set.

The data section should include two tables. The first table should be for the solubility tests. This table should include **(a)** the elements involved in the test, **(b)** observations for each test, and **(c)** solubility of each element. Make sure the observations in your lab report match those that you wrote in your lab notebook during lab. The second table should be for the reactivity tests and should include **(a)** the elements involved in the test **(b)** observations and **(c)** results of the test, both in terms of reactivity and electronegativity.

The calculation section should have the six balanced equations for the reactions being studied based on your electronegativity data. If no reaction occurred write "no reaction" as the product. The balanced equations should be in one of the following formats and either typed or written in ink. The counter ions, Na^+ and K^+, are spectators in the reaction and need not be included.

$$2X^- + Y_2 \rightarrow X_2 + 2Y^-$$

or

$$2X^- + Y_2 \rightarrow \text{no reaction}$$

The conclusion should be several well-developed paragraphs that include the following: a discussion of the results, including a list of the halogens in order of electronegativity and identification of the most electronegative element, with support from the data section. Your conclusions should also include a discussion of the agreement between your results and the prelab predictions. An explanation of how you decided which reactions occurred and which did not should be included. A final discussion should focus on what the solubility tests indicated. Finally, include a discussion of any errors in the experiment.

Answer the following questions:

1) This experiment uses a qualitative colorimetric process. What is meant by this statement?

2) All of the halogens, with the exception of fluorine, make strong acids. Using electronegativity, explain why hydrofluoric acid is a weak acid.

Experiment 12
Laboratory Preparation

___

Name: _____ Date: _____

Instructor: _____ Sec. #: _____

___

Show all work for full credit.

1) Read the background, procedure, and report sections of the lab experiment carefully and develop a hypothesis of what information you expect to gain from the completion of the lab experiment.

2) Create any and all tables you might need to collect data during the experiment and then transfer those tables into your lab manual for use during the lab.

Experiment 12
Prelaboratory Assignment

Name: _____ Date: _____

Instructor: _____ Sec. #: _____

Show all work for full credit.

1) When you go into your laboratory to carry out the halogen displacement reactions, you already know from lecture that the order of reactivity, and electronegativity, is $Cl_2 > Br_2 > I_2$, so you are able to predict which results to expect from the various combinations of halogen and halide salt.

Let us suppose, though, that you are a student with a color perception disorder, so you see the color of the halogens differently from other students. And suppose the halogen samples aren't identified for you, but are given as unknowns, labeled X_2, Y_2, and Z_2 for the halogens, and X^-, Y^-, and Z^- for the corresponding halides. You are to identify the relative reactivity of the unknowns, and hence their identity, by mixing different combinations of halogen and halide and identifying the halogen product from the color of the hexane layer after extraction.

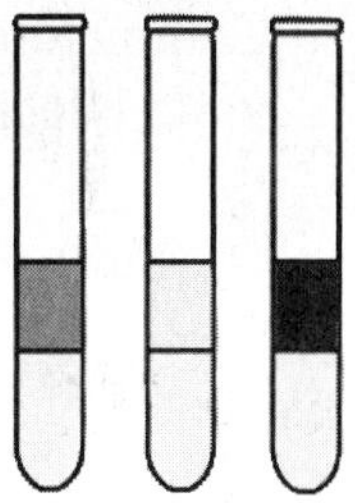

As a start, you first add the free halogens to your test tube, extract into hexane, and get the results shown in the figure to the right.

Next, you mix the following combinations of reagents and record the resulting color of the hexane in your test tubes, giving the results shown in the table below.

Z_2 X_2 Y_2

Reagent Combination	Resulting Color
$Y_2 + Z^-$	Red
$X_2 + Y^-$	Blue
$Z_2 + X^-$	Yellow

Now, rank the unknowns from most reactive (1) to least reactive (3).

Z_2 _____
Y_2 _____
X_2 _____

Predict the color that would result from the following reagent mixtures:

$Z_2 + Y^-$ _____
$X_2 + Z^-$ _____
$Y_2 + X^-$ _____

Now that you have identified the relative reactivities (and electronegativities) of the unknowns, match them to their correct halogen identities.

Z_2 _____
Y_2 _____
X_2 _____

2) Complete and balance the following reaction as a "molecular" equation. If no reaction occurs, write NR in place of giving products.

$$NaI + Cl_2 \rightarrow$$

Complete and balance the following reaction as a "net ionic" equation. If no reaction occurs, write NR in place of giving products.

$$NaBr + Cl_2 \rightarrow$$

3) Draw the structures of hexane and water below and describe why each is or is not a polar molecule.

Experiment 13
Building Molecular Models

Introduction

In any subject, there is always a language to learn. In chemistry, the language uses elements as the alphabet and compounds as the words. Reactions will eventually be the sentences, but for now, you will need to learn the proper way to write the words and then to develop a mental picture of the structures represented by these words. The use of Lewis structures gives us a step-by-step way to construct molecular models of any species: covalent or ionic. Lewis structures are based on the idea that every atom is most stable when its electron configuration matches that of the nearest noble gas. This is often called the *octet rule* because many common atoms are most stable when they have eight electrons in their outermost or valence shell.

Because Lewis structures are conceptualized at the molecular level, this experiment will be used as a workshop to practice drawing and visualizing molecular structures. In the lab itself, you will complete a worksheet that requires you to draw and analyze models of several inorganic compounds. Additionally, you will use a protractor to determine the bond angles of the resulting molecular structures.

Background

The premise behind Lewis structures is the octet rule: All atoms would like to be surrounded with an octet of electrons. Of course, there are, some exceptions: Very small atoms (H, Be, and B) have less than an octet, and some main group atoms with low-energy *d* orbitals (P, S, Cl, Br, and I) may have more than an octet. This is especially true when these atoms are central atoms and combined with highly electronegative atoms.

Drawing correct Lewis structures takes practice, but the process can be simplified by following a series of steps:

Step 1:   Count all the valence electrons for each atom. Add or subtract electrons if the structure is an anion or cation, respectively.

Step 2:   Determine which atoms are bonded to one another. To draw the skeleton of the molecule, select the least electronegative atom as the central atom, and surround it by the other atoms. Don't forget that H cannot be between other atoms because it can form only one bond.

Step 3:   Connect the atoms with a pair of electrons for each bond. Subtract the number of bonding electrons from the total number of valence electrons.

Step 4:   Add electron pairs to complete octets for all peripheral atoms attached to the central atom. Beware of hydrogen–hydrogen *never* has more than one bond or one pair of electrons

Step 5:   Place remaining electrons on the central atom, usually in pairs. The octet rule may be exceeded for P, S, Cl, Br, I, etc. if they are the central atom, known at the expanded octet. Atoms in the third period of the periodic table or below have empty orbitals that will allow them to accommodate these expanded octets.

Step 6:   The central atom must follow the octet rule. To verify the central atom has 8 electrons surrounding it, the total number of valence electrons should be counted. If there are 2 less electrons, one double bond should be added. Each bond adds 2 electrons. If there are 4 less electrons, a triple bond will be added. This will be done by moving electron pairs from one or more of the peripheral atoms to achieve the octet. The placement of these bonds is determined by the formation of the most stable resonance structures.

Step 7:   Look for resonance structures by rearranging bonds. The structure with the lowest total formal charges will be the most likely form to be found in nature. (See below for explanation.)

Step 8:   Assess your final structure. Make sure you have used all of the atoms and all of the valence electrons you calculated in Step 1.

Drawing Lewis Structures

Let's look at an example of how this works using a real molecule. Consider the molecule most responsible for the greenhouse effect, carbon dioxide (CO_2).

To draw the Lewis structure:

Step 1:   Count all the valence electrons for each atom:

1 carbon atom × 4 valence electrons = 4 electrons

2 oxygen atoms × 6 valence electrons = 12 electrons

Total = 16 electrons

Step 2: Determine which atoms are bonded to one another. Generally, the least electronegative atom is the central atom. However, if the only choice is between a more electronegative atom and hydrogen, the more electronegative atom will be the central atom (e.g., in water). Hydrogen NEVER makes more than one bond and thus can NEVER be the central atom.

For CO_2, carbon is the less electronegative atom, so it should be the central atom.

$$O : C : O$$

Step 3: Connect each atom with a single pair of electrons or single bond: (16 valence electrons – 4 bonding electrons = 12 electrons left).

$$O{-}C{-}O$$

Step 4: Add electron pairs to peripheral atoms for octets:

$$:\ddot{O}{-}C{-}\ddot{O}:$$

Step 5: No electrons are left over, but the central atom doesn't have an octet! Keep in mind that carbon must always have four bonds; therefore, it shouldn't have any lone pairs.

Step 6: Move electrons from peripheral atoms, forming double bonds, to give the central atom an octet:

$$\ddot{O}{=}C{=}\ddot{O}$$

See the next step for the reasoning on the placement of these double bonds.

Step 7: Look for resonance structures and identify the one with the smallest formal charges:

$$\ddot{O}{=}C{=}\ddot{O} \quad \rightleftharpoons \quad :\ddot{O}{-}C{\equiv}O: \quad \rightleftharpoons \quad :O{\equiv}C{-}\ddot{O}:$$

$$\;0\qquad 0\qquad 0 \qquad\qquad\;\; -1\qquad 0\qquad +1 \qquad\qquad\;\; +1\qquad 0\qquad -1$$

For some molecules, more than one structure can be drawn. Note that a Lewis structure for carbon dioxide can be written using a carbon–oxygen single bond on one side and carbon–oxygen triple bond on the other. How can these two possibilities be distinguished? How can the most important structure be chosen, or are they all equally likely? When several structures can be drawn, they are called resonance structures, which are discussed in more detail below.

Step 8: $$\ddot{O}{=}C{=}\ddot{O}$$

Experiment 13

The original formula called for one carbon and two oxygens, check. The total valence electrons were 16, and 16 have been used, check. This is the lowest energy (most stable) structure based on formal charges, check. This is a good structure. 😊

Resonance Structures

In resonance structures, all the atoms are in the same relative position to one another, but the distribution of electrons around them is different. To evaluate the importance of each structure, the formal charge on each atom must be determined.

Formal Charge

Formal charge is a somewhat arbitrary way of describing how many electrons an atom seems to have in a particular compound. Electron pairs in bonds between atoms are assumed to be shared equally between the two atoms. Nonbonding electron pairs are counted as belonging to the atom on which they reside. This can be put into an equation:

$$\text{formal charge} = \#\text{valence e}^-\text{s} - (\#\text{nonbonding e}^- + \tfrac{1}{2}\text{bonding e}^-)$$

or

$$\text{formal charge} = \#\text{valence e}^-\text{s} - (\#\text{nonbonding e}^- + \#\text{ of bonds})$$

The most stable resonance structure is the one in which:

- There is a minimum number of formal charges

- If there are formal charges, like charges are separated

- Negative formal charges are on the more electronegative atoms and positive formal charges are on the less electronegative atoms

For the CO_2 structure with two double bonds, the formal charges can be calculated as follows:

Oxygens: formal charge = $6 - (4 + 1/2(4)) = 0$

Carbon: formal charge = $4 - (0 + 1/2(8)) = 0$

For the CO_2 structure with a single and triple bond:

Oxygen (single): formal charge = $6 - (6 + 1/2(2)) = -1$

Oxygen (triple): formal charge = $6 - (4 + 1/2(6)) = +1$

Carbon: formal charge = $4 - (0 + 1/2(8)) = 0$

So, while both structures work as Lewis structures, the one that results in zero formal charges for any of the atoms is more stable and thus more likely to exist in nature than the one having charges on the two oxygen atoms.

Oxidation Numbers

Formal charges need to be distinguished from oxidation numbers (which can also be determined from Lewis structures). Oxidation numbers are used to indicate whether a molecule is neutral, electron rich, or electron poor. These rules are hierarchical and must be applied in the order written. A more complete set of rules for determining oxidation numbers are found in your textbook. Only a short summary of some of these rules is given here:

- The oxidation number for an element in its elemental form is 0 (true for isolated atoms and for molecular elements, e.g., Cl_2, S_8, and P_4).

- The oxidation number of a monatomic ion is the same as its charge (e.g., the oxidation number of $Na^+ = +1$, and that of S^{2-} is -2).

- In binary compounds, the element with greater electronegativity is assigned a negative oxidation number equal to its charge if it is found in simple ionic compounds. (E.g., in the compound PCl_3, the chlorine is more electronegative than the phosphorus. In simple ionic compounds Cl has an ionic charge of $1-$, so its oxidation state in PCl_3 is, -1.)

- The sum of the oxidation numbers is zero for an electrically neutral compound and equals the overall charge for an ionic species.

- Alkali metals exhibit only an oxidation state of $+1$ in compounds.

- Alkaline earth metals exhibit only an oxidation state of $+2$ in compounds.

Once Lewis structures are drawn successfully, they can be used to predict the electron cloud geometry, molecular shape, and polarity of molecules and ions. For a thorough discussion, refer to your textbook. In particular, look at the three-dimensional representations for all the geometries and shapes.

Electron Cloud Geometry and Molecular Geometry

The electron cloud geometry around a central atom is determined by the number of electron groups surrounding it. Each set (2, 3, 4, 5, and 6) has a different name and arrangement in three-dimensional space. Electron clouds, all being negative, are most stable when separated as far from one another as possible. This is called the *valence shell electron pair repulsion theory* (VSEPR). While electron cloud geometry describes the orientation of the electrons around the central atom, the molecular geometry describes the arrangement of peripheral atoms. While the Lewis structure itself tells us nothing about the three dimensional structure of the molecule, both the electron cloud geometry and the molecular geometry depend on the construction of a good Lewis structure to make their determinations of atom and electron placements within the molecule.

The Experiment

In the lab, you will be presented with six molecular models as unknowns. It will be your job to name them. You will also be asked to determine their electron cloud and molecular geometries by measuring their bond angles using a protractor. A worksheet is provided containing other questions that should be completed for each of the molecules. You should make five additional copies of the worksheet to use during class. These worksheets will then be used as the data section of your lab report.

Procedure

SAFETY NOTES: Since no chemicals will be used today, goggles are not required.

GENERAL INSTRUCTIONS: Each student works independently in this lab. Print out six worksheets to use for data collection during the experiment.

Part I: Identifying Lewis Structures

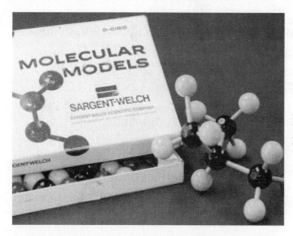

Draw a good **Lewis Structure** for each of the six molecular models present.

Part II: Analyzing Models

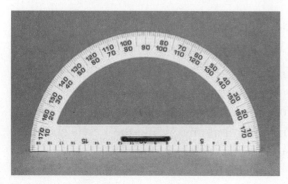

Collect a **protractor**.

Bond Angles

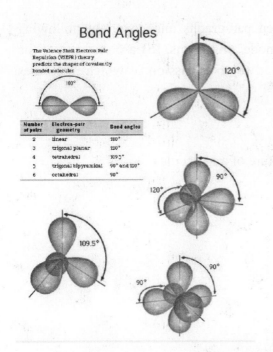

The Valence Shell Electron Pair Repulsion (VSEPR) theory predicts the shapes of covalently bonded molecules.

Number of pairs	Electron-pair geometry	Bond angles
2	linear	180°
3	trigonal planar	120°
4	tetrahedral	109.5°
5	trigonal bipyramidal	90° and 120°
6	octahedral	90°

Use the protractor to measure the angles between the bonds of the molecular models and record them on your worksheet.

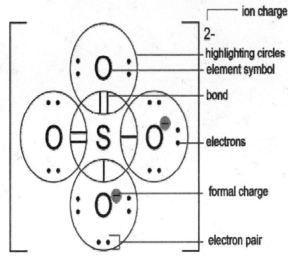

ion charge

2-

highlighting circles

element symbol

bond

electrons

formal charge

electron pair

For each Lewis Structure you have drawn, answer the remaining questions posed by the **worksheet**.

Report Contents and Questions

The purpose should be several well-constructed sentences dealing with the goals of this experiment and what would be required for the experiment to be considered a success. The procedure section should cite the lab manual and also note any changes made to the procedure.

The data section for this experiment should include the six worksheets that you completed during the course of the laboratory experiment.

The calculation section should include sample calculations for determining the formal charges and molecular weight (MW) for each molecule.

The conclusion section should be several well-developed paragraphs and cover the following information: **(a)** identification of each of the molecular model unknowns, **(b)** a discussion of the polarity of each molecule and how it was determined, and **(c)** a discussion of which molecules have resonance forms and why.

Answer the following questions:

1) What physical forces create the three-dimensional structure of a molecule?

2) What is the purpose in calculating formal charges?

Experiment 13
Laboratory Preparation

Name: _____ Date: _____

Instructor: _____ Sec. #: _____

Show all work for full credit.

1) Read the background, procedure, and report sections of the lab experiment carefully and develop a hypothesis of what information you expect to gain from the completion of the lab experiment.

2) Create any and all tables you might need to collect data during the experiment and then transfer those tables into your lab manual for use during the lab.

Experiment 13
Prelaboratory Assignment

Name: _____ Date: _____

Instructor: _____ Sec. #: _____

Show all work for full credit.

1) Below are three resonance structures for the thiocyanate ion.

Structure 1   Structure 2   Structure 3

$$[:\ddot{N}-C\equiv S:]^- \longleftrightarrow [\ddot{N}=C=\ddot{S}:]^- \longleftrightarrow [:N\equiv C-\ddot{S}:]^-$$

F   A   C   H   D   B   I   G   E

Identify the formal charge on the atoms indicated by the letters below.

A _____

C _____

E _____

H _____

Which of the three structures is the most stable?

2) For each of the molecules or ions below, draw its Lewis structure and then answer the question based on that structure. (The central atom in each case is the atom with the lowest electronegativity.)

How many valence electrons are in the following? (Remember to add an electron for each negative charge or to subtract one for each positive charge.)

BeF_2 PF_5 $AsCl_3$

How many lone pairs of electrons are on the central atom of the following?

PCl_3 ClF_3 SF_4

What is the electron cloud geometry of the following?

ClO_3^- IF_5 BrF_5

What is (are) the bond angle(s) at the central atom of the following?

H_3O^+ ClF_5 BCl_3

3) The following is a Lewis structure of the essential amino acid **tryptophan**. Tryptophan is one of the 20 amino acids found in proteins but is also a precursor of the neurotransmitter **serotonin**.

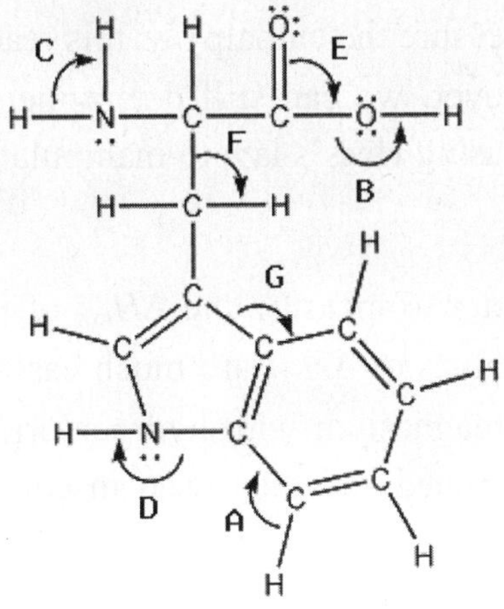

Tryptophan

Estimate the values of the bond angles in the tryptophan structure indicated by the following letters:

A _____

G _____

D _____

E _____

Experiment 14
Calorimetry and Hess's Law

Introduction

It is likely that you have started discussing the concepts of thermodynamics in class. In this lab, we will illustrate some of those concepts. But why in the world do you need to know about processes like calorimetry and Hess's law?

The purpose of this lab is to determine the enthalpy of reaction for the burning of one mole of magnesium in oxygen:

$$Mg(s) + \frac{1}{2}O_2(g) \rightarrow MgO(g) \quad \Delta H_{rxn} = \text{experimental goal}$$

The reaction is quite exothermic. Therefore, trying to measure the enthalpy of this reaction directly would be difficult and possibly even dangerous. However, we can still determine the ΔH_{rxn} by using calorimetry to collect experimental data and then using Hess's law to manipulate that data to yield the answer we need.

We will construct and use a constant pressure calorimeter to measure the ΔH_{rxn} of two reactions related by their reagents to our target reaction. These values of ΔH_{rxn} are much easier to measure because they are much less exothermic. The ΔH_{rxn} of magnesium with hydrochloric acid and of magnesium oxide with hydrochloric acid will be determined and then used in conjunction with Hess's law to calculate the ΔH_{rxn} for our target reaction.

Hess's Law Reactions

$$Mg(s) + 2HCl(aq) \rightarrow MgCl_2(aq) + H_2(g) \quad \Delta H_{rxn} = \text{experimental B}$$
$$MgO(s) + 2HCl(aq) \rightarrow MgCl_2(aq) + H_2O(l) \quad \Delta H_{rxn} = \text{experimental C}$$
$$2H_2(g) + O_2(g) \rightarrow 2H_2O(l) \quad \Delta H_F = -285.840 \text{ kJ/mol}$$

Background

In studying chemical processes, attention usually centers upon the properties of the substances involved. However, the energy changes of these reactions are also very important. A *system* is that region of the universe under consideration. The *surroundings* are everything other than the system. Thus, the state of the system is specified by a number of variables, including temperature, pressure, volume, and chemical composition. A more extensive discussion of these energy relationships can be found in your textbook.

When a system undergoes any chemical or physical changes, the *first law of thermodynamics* (conservation of energy) requires that the accompanying change in the system's *internal energy* (ΔE) is equal to *heat* (q) plus *work* (w):

$$\Delta E = q + w \qquad \text{Equation 1}$$

Thus, the energy of a system increases when heat is added (q is positive) and/or work is done on the system (w is positive). For systems under constant pressure (note subscript p), all work being done is based on the change in volume ($w = -P\Delta V$), and Equation 1 becomes:

$$\Delta E = q_p - P\Delta V \qquad \text{Equation 2}$$

The value for the heat (per mole) absorbed or given off by a system at constant pressure (q_p) is called the *heat of reaction* or *enthalpy of reaction* (ΔH). By substituting ΔH for q_p and rearranging, Equation 2 becomes:

$$\Delta H = \Delta E + P\Delta V \qquad \text{Equation 3}$$

If there is no change in volume (no ΔV), then ΔH is the change in energy of the system at constant pressure. Therefore, ΔH is a useful quantity to measure for reactions in solution (at constant pressure) where no gaseous products are formed (no change in volume). This describes a lot of chemistry. (NOTE: Although one of the reactions we are using in this experiment does produce a gas, the change in volume is negligible. So, we will maintain our assumption of constant volume. As with all assumptions, we should acknowledge it as a possible source of error in the conclusion.)

The amount of heat (q) necessary to raise the temperature of a system is an extensive property. That is, it depends on the amount of material, as well as what the material is and how much the temperature changes. This can be expressed in two ways:

$$q = n\, C_m\, \Delta T \qquad \text{Equation 4}$$

where n is the number of moles of material, C_m is the molar heat capacity, and ΔT is the change in temperature (either in K or °C; since it's a *change*, numerically it doesn't matter, but your units for C_m should be matched). Alternatively, this can be expressed for the mass of the material as:

$$q = m\, c_p\, \Delta T \qquad \text{Equation 5}$$

where m is the mass of the material in grams, c_p is called the specific heat of the material, and ΔT is the change in the temperature (either in K or °C). If the conditions are restricted to constant pressure, these expressions become:

$$q_p = n\, C_m\, \Delta T \quad \text{and} \quad q_p = m\, c_p\, \Delta T$$

The formal definitions of the constants described above are: *Molar heat capacity* (C_m) is the amount of heat required to raise the temperature of one mole of material by one degree Kelvin or Celsius. *Specific heat* (c) is the amount of heat required to raise the temperature of one gram of material by one degree Kelvin or Celsius. The value of C_m or c_p depends on the identity of the material, its state (gas, liquid, solid), and its temperature. Below is a table of the values of these constants for various states and temperatures of water. Note that for most of the range of liquid water temperatures, the value does remain constant to two significant figures.

TABLE 14.1 Molar Heat Capacities and Specific Heats for Water

State	Temperature (°C)	Temperature (K)	Heat Capacity (J/mol × K)	Specific Heat (J/g × K)	Specific Heat (cal/g × K)
Solid	−34.0	239.2	33.30	1.846	0.4416
Solid	−2.0	271.2	37.78	2.100	0.5024
Liquid	0.0	273.2	75.86	4.218	1.0070
Liquid	25.0	298.2	75.23	4.180	0.9983
Liquid	100.0	373.2	75.90	4.216	1.0070
Gas	110.0	383.2	36.28	2.010	0.4810

An older (English system) unit used to express heat is the *calorie*. It is defined as the amount of heat necessary to raise 1 g of water from 14.58 °C to 15.58 °C at one atmosphere pressure. Obviously, since much work is done in aqueous solutions, this is still a useful unit, even though it is not an SI unit. (NOTE: The food Calorie, abbreviated Cal, is actually 1000 calories, or one kilocalorie.)

To determine the state function, enthalpy (ΔH), the heat (q_P) of a particular reaction, is divided by the number of moles of material involved in the reaction.

$$\Delta H = \frac{q_P}{n} \qquad \text{Equation 6}$$

The Calorimeter Cup

The term *calorimetry* refers to the measurement of heat released or absorbed during a chemical or physical process. The ideal calorimeter is well insulated so that its contents do not gain or lose heat to the surroundings and is constructed of a material of low heat capacity so that only a small amount of heat is exchanged between the contents and the calorimeter. For many processes, a simple unsealed, insulated cup can be used as a calorimeter because it has a low heat capacity and excellent insulating properties. Constant pressure on the reaction is maintained by the atmosphere.

In reality, no calorimeter is ideal. Thus, to obtain reliable results, the calorimeter must be calibrated to determine how much heat is exchanged with the calorimeter cup (and/or other surroundings). This correction factor is called the *calorimeter constant* (heat capacity of the calorimeter in J/K) and fits into an equation similar to the ones above. The temperature change of the cup can be assumed to be the same as that of the solution in it. Since the ΔT of the cup will change with each experiment, it is useful to have a calorimeter constant that can be used in all experiments as follows:

$$q_{cal} = C_{cup}\Delta T \qquad \text{Equation 7}$$

The heat capacity of the calorimeter cup (C_{cup}) is determined by performing a separate experiment in which no chemical reaction takes place. Instead, hot water is added to the cold water in the calorimeter. The amount of heat lost by the hot water (q_{hw}) must be equal to the heat gained by the cold water (q_{cw}) plus the heat gained by the calorimeter cup (q_{cal}). (We can't lose energy, so it only has two places to go.) This can be written as:

$$-q_{hw} = q_{cw} + q_{cal} \qquad \text{Equation 8}$$

Note that the signs of these amounts of heat are opposite because the hot water is losing energy, and the cold water and cup are gaining energy.

The amount of heat (q_{cal}) absorbed by the cup is easily found, since q_{hw} and q_{cw} can be calculated using Equation 5. To determine q_{hw}, substitute the mass of the hot water, the specific heat of water (Table 14.1), and the temperature change for the hot water (ΔT) into the equation. The q_{cw} is found the same way. Once q_{cal} is found, C_{cup} can be calculated from Equation 7.

To clarify the principles of calorimetry, consider the following hypothetical experiment. Using calorimetry methods, a determination of the heat of reaction of NaOH with HCl is made.

$$NaOH(aq) + HCl(aq) \rightarrow NaCl(aq) + H_2O(l)$$

The heat capacity for the calorimeter cup is given as 6.80 J/K. In the experiment, 75 mL (75 g, assuming a density of 1.0 g/mL) of a 0.6 M HCl solution is placed in the calorimeter and allowed to equilibrate to room temperature. Then 75 mL (75 g) of a 0.6 M NaOH is added to the calorimeter with stirring. Temperature readings are taken before and after addition of the NaOH for 10 min. Plotting temperature versus time gives the following plot.

Time (sec)	Temp. (°C)
0	22.8
30	22.8
60	22.9
90	23.6
120	28.2
150	27.8
180	27
210	26
240	25.5
270	24.8
300	24.1
330	23.7
360	23.6

Temperature vs. Time

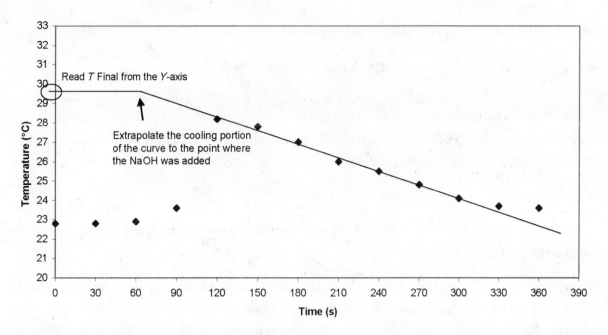

The initial temperature (T_i) of the acid in the cup is 22.8 °C. Temperature readings are started one minute before the addition of the second substance. Upon addition of NaOH, the acid temperature increases to a maximum of 28.2 °C in one minute and then slowly decreases. The precise determination of T_f is more complicated, since heat exchange is occurring between the contents and the cup during and after the reaction. Evaluation of T_f is accomplished by extrapolation as shown above. This gives a theoretical temperature representing the temperature obtained if instantaneous equilibrium was achieved within the system. By extrapolation, T_f = 29.7 °C. The heat capacity for the calorimeter cup is given as 6.80 J/K.

Calculations

NOTE: Since these calculations require only the use of ΔT, is not necessary to convert all Celsius temperatures to Kelvin. (Convince yourself this is true!)

$$\Delta T = T_f - T_i = 29.7\,°C - 22.8\,°C = 6.9\,°C \text{ (also 6.9 K if a } \Delta T)$$

To determine the heat of reaction, use Equation 5 and Table 14.1. (Assume that the heat capacity for a dilute solution of acid in water is very similar to that of water.)

$$q_{HCl} = m\,c_p\,\Delta T = 75.0\,g \times 4.180\,J/g\,K \times 6.9\,K$$

$$q_{HCl} = 2163\,J = 2.2\,kJ$$

We also need to calculate the heat absorbed by the calorimeter cup.

$$q_{cal} = C_{cup} \times \Delta T = 6.80\,J/K \times 6.9\,K$$

$$q_{cal} = 47\,J$$

The final heat of reaction value is calculated as follows:

$$q_{rxn} = q_{HCl} + q_{cal} = 2163 \text{ J} + 46.9 \text{ J}$$

$$q_{rxn} = 2210 \text{ J} = 2.2 \text{ kJ}$$

Note that the calorimeter constant is very small compared to the q_{HCl} and did not affect the final result when expressed to two significant figures.

Hess's Law

If a reaction is carried out in a series of steps, ΔH for the reaction is equal to the sum of the enthalpy changes for the individual steps.

The overall enthalpy change for a process is independent of the number of steps or the particular nature of the path by which the reaction is carried out. Thus, we can use information tabulated for a relatively small number of reactions to calculate ΔH for a large number of different reactions.

In this experiment, we want to know the heat of reaction for burning magnesium metal in oxygen:

$$2Mg(s) + O_2(g) \rightarrow 2MgO(s)$$

This is a very exothermic reaction whose ΔH of reaction would be extremely difficult to measure using our calorimetry set-up. However, by experimentally measuring the heat of reaction for two other reactions:

$$Mg(s) + 2HCl(aq) \rightarrow MgCl_2(aq) + H_2(g)$$
$$\Delta H_{rxn} = \text{(Experiment B)}$$

and

$$MgO(s) + 2HCl(aq) \rightarrow MgCl_2(aq) + H_2O(l)$$
$$\Delta H_{rxn} = \text{(Experiment C)}$$

and using the heat of formation reaction for water:

$$2H_2(g) + O_2(g) \rightarrow 2H_2O(l)$$
$$\Delta H_f = -285.840 \text{ kJ/mol}$$

we can use Hess's law to calculate the heat of reaction for magnesium burning in oxygen. Refer to your textbook for further explanation on how to use Hess's law.

Procedure

SAFETY NOTES: The thermometers are fragile! If one is broken, inform your instructor immediately. Thoroughly wash off any acid or base that comes in contact with your skin. Acids and bases can cause burns!

GENERAL INSTRUCTIONS: Students will work in pairs, with one person swirling the cup and reading the thermometer while the other keeps time and records the data in both lab notebooks.

Experiment A: Determination of the heat capacity of the calorimeter (calorimeter constant)

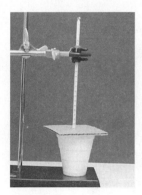

The calorimeter consists of an insulated cup covered with a cardboard lid. The lid has a hole to accommodate the thermometer (see picture). A clamp may be used to support the thermometer. After assembling the calorimeter, make sure there is room to swirl the calorimeter without bumping the thermometer.

In *each* part of the experiment, measure the volume of solution put in the calorimeter with a graduated cylinder. Record these values to be used in calculations later. Allow the calorimeter and solution to come to room temperature. Equilibrium is reached when the temperature remains constant for two to three temperature readings at 30 sec intervals. Once equilibrium is reached, you will add the second material to the calorimeter and continue recording the temperature at 30 sec intervals for the amount of time specified. This will provide sufficient data to determine T_f by extrapolation. The calorimeter must be swirled constantly during the temperature reading phase of the experiment to ensure complete mixing of the contents.

Place 150 mL of water in a 400 mL beaker and heat to 50 to 60 °C.

While the water is heating, measure out 50 mL of cold water in a graduated cylinder and place it in your calorimeter cup.

Monitor the temperature of the cold water in the cup until it remains constant for two to three readings at 30 sec intervals.

Measure 50 mL of your hot water into a graduated cylinder.

Return the thermometer to the calorimeter cup and record the temperature for three readings at 30 sec intervals.

Using the thermometer record the temperature of the hot water.

Lift the lid of the calorimeter and pour the 50 mL of hot water in, mixing continuously.

Record the temperature every 30 sec until 10 min have elapsed.

Dispose of the water in the calorimeter down the sink and repeat Part A of the experiment.

Experiment B: Determination of the Heat of Reaction of Mg(s) in HCl

Make sure your calorimeter cup is clean and dry and then add 100 mL of 1.0 M HCl to the cup.

Weigh out four strips of magnesium metal to the closest 0.001 g. (You can choose to use any weighboat, watch glass etc. to contain the magnesium for weighing.)

Place the thermometer in the calorimeter cup and record the temperature for three readings at 30 sec intervals.

Lift the lid of the calorimeter and drop the pieces of magnesium in, mixing continuously.

Record the temperature every 30 sec until 10 min have elapsed.

Dispose of the solution in the waste jar in the hood and rinse and dry the calorimeter cup thoroughly.

Experiment C: Determination of the Heat of Reaction of MgO in HCl

Add 100 mL of 1.0 M HCl to your calorimeter cup. (Make sure the cup was thoroughly rinsed and dried from the previous experiment.)

Weigh out ~0.50 g of magnesium oxide to the closest 0.001 g. (You can choose to use any weighboat, watch glass etc. to contain the magnesium for weighing.) NOTE: MgO is a powder that is vulnerable to being blown everywhere, so be careful when transferring it to and from containers.

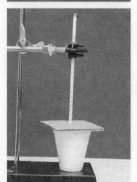

Place the thermometer in the calorimeter cup and record the temperature for three readings at 30 sec intervals.

Lift the lid of the calorimeter and drop the magnesium oxide in, mixing continuously. (IMPORTANT: If you do not stir at this point, the MgO will drop to the bottom of the cup, and the reaction will not be complete.)

Record the temperature every 30 sec until 10 min have elapsed.

Dispose of the solution in the waste jar in the hood and rinse and dry the calorimeter cup thoroughly.

Report Contents and Questions

The purpose should be several well-constructed sentences that describe the goals of the experiment and the criteria for determining the success or failure of the experiment. The procedure section should reference the lab manual and include any changes that were made to the procedure during lab.

The data section for this lab is very large. A fill-in-the-blank worksheet has been provided, but you should create your own data table for your lab report; this can be in any format. You can put all the information in one table or split it by experimental section; just make sure the table includes all of the information found in the worksheet.

Worksheet for Calorimetry

	Trial 1	Trial 2
Part A		
Exact volume of cold water (to the nearest 0.1 mL)	50 mL	50 mL
Temperature of cold water (in cup)	20°C	21°C
Exact volume of hot water (to the nearest 0.1 mL)	50 mL	50 mL
Temperature of hot water (in cylinder)	49°C	50.1°C

Part A Calculations:

	Trial 1	Trial 2
Mass of cold water (assume density = 1.00 g/mL)	50 g	50 g
T_f from graph by extrapolation	30.9°C	30.8°C
ΔT_{hw} for hot water	-18.1	-19.3
ΔT_{cw} for cold water	10.9	10.8
q_{hw} for hot water (use $q_{hw} = m\,c\Delta T_{hw}$)	3786.52	4037.56
q_{cw} for cold water (use $q_{cw} = m\,c\,\Delta T_{cw}$)	2280.78	2259.36

Weigh boat = 4.5330 g   strips = 0.1635 g
WB + 4 strips = 4.6965 g

q_{cal} for the cup (use $|q_{hw}| = |q_{cw}| + q_{cal}$, solve for q_{cal})

C_{cup} for the cup (use $q_{cal} = C_{cup} \Delta T$)

Which ΔT should be used?

150624   1780.2
138.19   181.65

Part B

Description of sample. Mg = Chrome   HCl = Clear

Exact volume of HCl (to the nearest 0.1 mL)   100 mL

Initial temperature of HCl (in cup)   19°C

Exact mass of Mg (to the nearest 0.001 g)   0.1635 g

Part B Calculations:

T_f from graph by extrapolation   25.8

Mass of HCl solution (use the density of water for HCl = 1.00 g/mL)   100 g

ΔT_{cw} for HCl   6.8

q_{HCl} for HCl solution (use $q_{cw} = m\,c\,\Delta DT_{HCl}$)   2845.16

q_{cal} for the cup (use $q_{cal} = C_{cup}\Delta T_{HCl}$)   1087.46

q_{rxn} (use $q_{rxn} = (q_{HCl} + q_{Cal})$)   3932.58

ΔH_{rxn} for Mg (use $\Delta H_{rxn} = q_{rxn}/n_{Mg}$)   642.98

Write the net reaction which took place in the cup.

Part C

Description of sample

Exact volume of HCl (to the nearest 0.1 mL)   100 mL

Temperature of HCl (in cup)   19.1°C

Exact mass of MgO (to the nearest 0.001 g)   0.5082 g

Part C Calculations:

T_f from graph by extrapolation   22.0

ΔT_{cw} for HCl   2.9

q_{HCl} for HCl solution (use $q_{cw} = m\,c\,\Delta T_{HCl}$)   1213.36

q_{cal} for the cup (use $q_{cal} = C_{cup}\,\Delta T_{HCl}$)   1929.24

q_{rxn} (use $q_{rxn} = (q_{HCl} + q_{cal})$)   3142.6

ΔH_{rxn} for MgO (use $\Delta H_{rxn} = q_{rxn}/n_{MgO}$)   6183.78

Write the net reaction that took place in the cup.

Most of the calculations are explained in the worksheet above. Recall, though, that the change in temperature (ΔT) is defined as: $\Delta T =$ **final temperature – initial temperature.** For each part (A, B, and C) include a table of some type with the time and temperature data gathered in lab. A temperature versus. time graph for each run should be included. To determine T_f, draw a trend line and extrapolate the line to the time the two substances were mixed (probably around 60 to 90 sec) draw a horizontal straight line from that point to the y-axis, and record the temperature at which the line and y-axis intercept.

Example Graph:

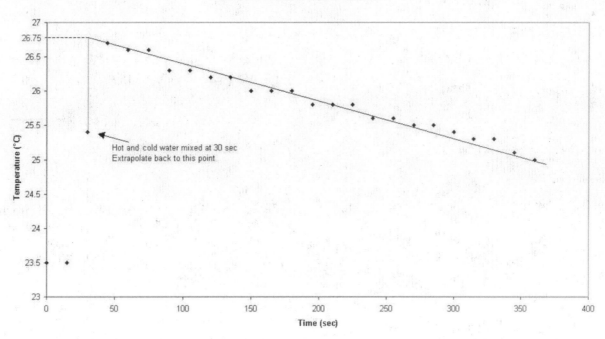

The calculation section should in include temperature versus time graphs for each part. All graphing guidelines apply here. Sample calculations should be included for the following: the heat capacity of the calorimeter, heat of reaction for both Parts B and C, and ΔH for both Parts B and C.

Using your experimental values of ΔH_{rxn} from Parts B and C and Hess's law, calculate the ΔH_{rxn} for

$$\text{Mg}(s) + \frac{1}{2}\text{O}_2(g) \rightarrow \text{MgO}(g) \quad \Delta H_{rxn} = \textbf{experimental goal}$$

Be sure to show the equations and setup used. Finally, compare your value with the literature value of ΔH_{rxn} and determine the percent error in your experiment.

The conclusion section for this experiment should be several well-developed paragraphs. These paragraphs should cover the following information:

1) What is the difference between the heat (q) of a reaction and the reaction enthalpy (ΔH_{rxn})?

2) Why is T_{final} taken from extrapolating back to the beginning of the experimental run rather than from the last temperature reading taken?

Experiment 14
Laboratory Preparation

Name: _____ Date: _____

Instructor: _____ Sec. #: _____

Show all work for full credit.

1) Read the background, procedure, and report sections of the lab experiment carefully and develop a hypothesis of what information you expect to gain from the completion of the lab experiment.

2) Create any and all tables you might need to collect data during the experiment and then transfer those tables into your lab manual for use during the lab.

Experiment 14
Prelaboratory Assignment

Name: _____ Date: _____

Instructor: _____ Sec. #: _____

Show all work for full credit.

1) Use the graph of temperature (°C) versus time (s) provided to answer the following questions:

   What is the initial temperature? What is the final temperature?

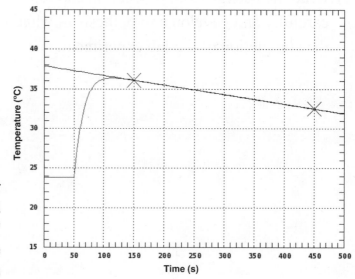

   The C_p for the calorimeter (including the HCl solution and the Mg) was calculated to be 0.247 kJ/°C. For the temperature curve shown above, 0.174 g Mg was reacted with HCl.

   What is the ΔH of the reaction?

2) To calibrate your calorimeter cup, you first put 45 mL of cold water in the cup and measure its temperature to be 21.7 °C. You then pour 59 mL of hot water, temperature = 50.1 °C, into the cup and measure the temperature every 30 sec over a 10 min period. You extrapolate this "cooling curve" back to the time of addition and find that the "final temperature" after mixing is 36.2 °C.

   What is the heat change of the hot water, q_{hw}?
   (Assume the density of the water is 1.00 g/mL and remember that the specific heat of water is 4.184 J/g-K or J/g-°C.)

What is the heat change of the cold water, q_{cw}?

What is the heat change of the calorimeter cup, q_{cup}?

What is the heat heat capacity of the calorimeter cup, C_{cup}?

3) You would like to measure the heat of neutralization of an acid with a base. You mix 225 mL each of 0.65 M HNO_3 and 0.65 M LiOH, both at a temperature of 18.1 °C in a calorimeter cup equilibrated to that same temperature. After following the temperature change for 10 min and extrapolating it back to the time of addition, you find that the "final temperature" after mixing was 22.3 °C. Previously, you measured the heat capacity of the calorimeter cup (C_{cup}) to be 47 J/°C.

Assuming that the density of the two solutions is 1.00 g/mL and that their specific heats are the same as water, 4.184 J/g °C, What is the heat absorbed by the two solutions, q_{soln}?

What is the heat absorbed by the calorimeter cup, q_{cup}?

What is the heat change for the neutralization reaction, q_{rxn}?

What is the ΔH for the neutralization?

Write the net ionic equation for this neutralization reaction, including the states of matter (i.e., $H^+ aq$) for the reactants and products.

4) Given the following data:

$2C_2H_6(g) + 7O_2(g) \rightarrow 4CO_2(g) + 6H_2O(l)$	$\Delta H = -3120$ kJ
$2H_2O(l) \rightarrow 2H_2(g) + O_2(g)$	$\Delta H = +572$ kJ
$C_2H_2(g) \rightarrow 2H_2(g) \rightarrow C_2H_6(g)$	$\Delta H = -312$ kJ

Find the ΔH of the following reaction:

$$4CO_2(g) + 2H_2O(l) \rightarrow 2C_2H_2(g) + 5O_2(g)$$

Experiment 15
Gas Laws and Air Bags

Introduction

Throughout this term, we have provided experiments that illustrate some of the more complex concepts of chemistry as you were learning about them in lecture. We have also presented you with evidence that each of the concepts or techniques covered was either valuable to your future as a chemist or relevant to your everyday life. This experiment on gas laws and airbags is both. Gas laws and the general concepts of gas volumes and pressures are vital for anyone deciding on a career in chemistry and even more important to those focusing on medicine. For example, oxygen saturation in the body is controlled by both internal and external pressures. Why is it harder to breathe at higher altitudes? What is hypoxia? What happens to a patient who receives too much oxygen? Not enough? And how do you as the doctor or nurse control the volume delivered? None of these questions can be answered without a firm understanding of the gas laws.

The experiment you will conduct today will not require you to administer oxygen to a patient (hopefully); rather, it focuses on the automobile airbag, another very useful device that depends on gas laws. Airbag deployment is chemically generated because mechanical deployment would be too slow. The explosive production of nitrogen gas from sodium azide fills the airbag in a fraction of a second. Gas laws allow the construction of a completely full but not overfull bag of gas that keeps the driver and passenger(s) of a car from serious injury during a collision.

To study how gas laws are used in the construction of airbags, you are going to design and build one. In addition to constructing the airbag, we are also going to revisit the use of a barometer to measure atmospheric pressure. Some barometers use a Vernier scale, so make sure you remember how to read this type of scale (See Experiment 2: A Submarine Adventure for a refresher).

Background

Though based on assumptions that gas molecules do not interact with each other and have negligible size, assumptions you will later learn are over-simplified, the ideal gas law is still incredibly useful in characterizing the properties of most gases. Mathematically, the ideal gas law is expressed by the equation $PV = nRT$, where P is the gas pressure, V is the volume in liters, n is the number of moles, T is the Kelvin temperature, and R is a constant.

The constant R is called the gas constant and takes on a number of different values depending on the units used for pressure. Note that although the units for pressure and volume can vary, the unit for temperature must always be in Kelvin.

Common Values of R
$8.314472 \: J \cdot K^{-1} \cdot mol^{-1}$
$0.0820574587 \: L \cdot atm \cdot K^{-1} \cdot mol^{-1}$
$8.20574587 \times 10^{-5} \: m^3 \cdot atm \cdot K^{-1} \cdot mol^{-1}$
$8.314472 \: L \cdot kPa \cdot K^{-1} \cdot mol^{-1}$
$62.3637 \: L \cdot mmHg \cdot K^{-1} \cdot mol^{-1}$
$62.3637 \: L \cdot Torr \cdot K^{-1} \cdot mol^{-1}$
$1.987 \: cal \cdot K^{-1} \cdot mol^{-1}$

In this experiment, you will use the ideal gas law to predict the volume of gas produced from a simple chemical reaction. This reaction will be used to create a mock automobile airbag.

Airbags

Airbags are safety devices found in most cars produced today. The reaction used in commercial airbags depends on sodium azide (NaN_3), a fairly toxic chemical. When activated, the airbag's sodium azide rapidly undergoes a decomposition reaction that generates sodium metal (Na) and nitrogen gas (N_2).

$$2NaN_3(s) \rightarrow 2Na(s) + 3N_2(g)$$

The nitrogen gas inflates the airbag to provide a safety cushion for the driver and passenger(s) in the car. The National Highway Traffic and Safety Administration says airbags deploy in 0.015 sec for high speed crashes and 0.025 sec for low speed crashes.

The Reaction

Since your laboratory is not equipped to work with chemicals as toxic as sodium azide, we will use sodium bicarbonate ($NaHCO_3$) to produce the gas. When sodium bicarbonate reacts with acetic acid (CH_3COOH), carbon dioxide (CO_2) and two other products are formed as shown below:

$$\mathbf{NaHCO_3(s) + CH_3COOH(aq) \rightarrow CH_3COONa(aq) + H_2O(l) + CO_2(g)}$$

The stoichiometry of this reaction is quite simple because all reactants and products are in a 1:1 molar ratio. Because the overall goal of this experiment is to design an airbag that inflates rapidly and fully, without wasting materials, you will need to convert molar amounts of reagents into gram and volume amounts to produce the precise volume of CO_2 that will fill your airbag.

Stoichiometry and Manipulating the Ideal Gas Law

To be able to predict the volume of the gas being produced, you will use the ideal gas law equation. The variables in the ideal gas law are volume, temperature, pressure, and the number of moles of gas. We know that our unknown in the equation is going to be the amount of gas produced. Therefore, we can rearrange the ideal gas law to solve for n:

$$n = \frac{PV}{RT}$$

Since we know R is a constant, all we have to measure is the volume of our bag, the atmospheric pressure, and the temperature of our surroundings.

For this experiment, we will use room temperature and pressure. A thermometer should be used to determine the room temperature and a barometer to measure the atmospheric pressure. With the temperature and pressure taken care of, all that remains is to determine the volume of your airbag. This can be done by filling the bag with water and then measuring the volume of water. The water also allows you to check the bag for leaks, which would ruin your experimental results.

By measuring the volume of gas needed and substituting that value along with room temperature and pressure, into the ideal gas law, you can predict the moles of gas required to fill the airbag. You can then use simple stoichiometry to calculate the exact amounts of reactants ($NaHCO_3$ and CH_3COOH) needed to produce the correct amount of CO_2 to fill the bag.

Example Problem

To further clarify the principles and calculations in this experiment, let's look at an example incorporating real data from an airbag with sodium azide. We need to find out how many grams of NaN_3 are needed to completely fill a 75.0 L airbag to a pressure of 1.30 atm at 25.0 °C with N_2. To find the mass of sodium azide required, we need to calculate the number of moles of N_2 needed.

$$n = \frac{PV}{RT} = \frac{(1.30\,\text{atm})(75.0\,\text{L})}{(0.08206\,\frac{\text{L·atm}}{\text{mol·K}})(298.15\,\text{K})} = 3.99\ \text{mol}\ N_2$$

Substituting these values into our equation as shown, we calculate that we need 3.99 mol of nitrogen gas. Knowing the number of moles of N_2 required, we can use the balanced equation given at the beginning of this discussion to determine the number of grams of NaN_3 needed.

$$3.99\ \text{mol}\ N_2\left(\frac{2\ \text{mol}\ NaN_3}{3\ \text{mol}\ N_2}\right)\left(\frac{65.02\ \text{g}\ NaN_2}{1\ \text{mol}\ NaN_2}\right) = 173\ \text{g}\ NaN_3$$

Therefore, we need ~173 g of sodium azide to produce 75.0 L of N_2. The calculations just completed are almost identical to those you will perform in this experiment.

Procedure

SAFETY NOTES: Acids can cause burns if they come in contact with your skin. Wash your hands immediately if you feel a burning or itching sensation. Use all necessary precautions when handling the acid container.

Part I: Calculating the Volume of the Airbag

Obtain one plastic airbag and determine its volume.

Fill your airbag as full as possible with water and use a graduated cylinder to calculate the volume of gas needed by determining the volume of water that filled the bag.

Part II: Testing Your Airbag

Use paper towels or Kimwipes to completely dry the inside of the bag.

Weigh out your calculated amount of sodium bicarbonate and add it to your airbag. Be sure to record the exact amount (in grams) you add.

little less NaCO3
little more acid

Measure out your calculated volume of 6.0 M acetic acid using a graduated cylinder. Record the volume to the nearest 0.1 mL. NOTE: Be very careful when handling the acid and wash your hands immediately if you get some on your skin. Acetic acid is not a strong acid, but it will still cause irritation if left in contact with your skin.

Test your airbag by quickly but carefully pouring your acetic acid into the airbag and then sealing it. Mix the ingredients completely by shaking and squishing the bag.

Once the reaction is complete, be sure to observe the fullness of the airbag and whether the reagents seem to be completely used up

Repeat the first six steps making adjustments to your calculated amounts of reagents if necessary. Dump the fluid remaining in the bag in the waste jar in the hood and rinse and dry your bag thoroughly. Be sure to show a successful airbag to your instructor or teaching/lab assistant before leaving.

140
+ 300
+ 300
− 50

weigh boat = 4.33 3.45

n = .028654769

V = 690 mL

4.7758M = .69C(.08)(243.45)

ml acetic acid

7.7207g CH3COOH

2.4072g NaCO3

188

Report Contents and Questions

The purpose should be in paragraph form and should describe the concepts of the lab and the criteria for a successful outcome for the experiment.

The procedure should reference the lab manual and note any changes made in the procedure.

The data section for this experiment should include a table like the one below. A balanced chemical equation for the reaction being studied should also be included.

Bag Dimensions	Bag Volume (L)	Moles CO_2 Needed	Moles $NaHCO_3$ Needed	Moles CH_3COOH Needed	Grams $NaHCO_3$ Needed	Volume of 6.0 M CH_3COOH Needed	Observations	Success or Failure

Room temperature: __293.3 K__ 20.3 C   Room pressure: __1 atm__

The column labeled success or failure should have a conclusion based on the results you observed for each airbag test.

The calculation section should include sample calculations for each column in the data table. Be sure to include conversions of units.

The conclusion section should be in paragraph format and discuss the success or failure of each airbag design. This should include the volume of each airbag and the amount of materials used. Along with the components of the airbag, you should include a discussion of the reasons for the success or failure of each bag. Also cite any problems or errors that may have occurred in the experiment and any additional trials that may have been needed.

Answer the following questions:

1) Is the gas produced in this experiment really an "ideal" gas? Explain.

2) The production of gas in this experiment is fairly slow, but its production in a car airbag is quite rapid. What is the reason for the difference in the two reactions?

Experiment 15
Laboratory Preparation

Name: _____ Date: _____

Instructor: _____ Sec. #: _____

Show all work for full credit.

1) Read the background, procedure, and report sections of the lab experiment carefully and develop a hypothesis of what information you expect to gain from the completion of the lab experiment.

2) Create any and all tables you might need to collect data during the experiment and then transfer those tables into your lab manual for use during the lab.

Experiment 15
Prelaboratory Assignment

Name: _____ Date: _____

Instructor: _____ Sec. #: _____

Show all work for full credit.

1) Airbags in cars are devices that inflate very rapidly (in about 30 ms) when there is a collision, creating a cushion that deflates slowly and reduces the impact of the collision on the driver and/or the passenger(s). The bag is inflated by a gas, which is generated by the rapid heating of sodium azide (NaN_3) to about 300 °C, causing it to decompose into sodium metal and nitrogen gas.

Complete and balance the equation for this reaction, indicating the states of the reactant ($NaN_3(s)$) and products ($Na(s)$ and $N_2(g)$).

Enough nitrogen must be generated in the bag to create a total pressure of 3.10 atm, which then drops as the bag slowly deflates. Assuming the volume of the bag is 58.0 L and the temperature in the car is 25 °C, calculate the molar quantity of N_2 that must be generated.

What molar quantity of sodium azide is required to generate this much nitrogen?

What mass of sodium azide is required?

The sodium metal produced in the reaction is highly reactive and potentially explosive, so it is removed in a reaction with excess $KNO_3(s)$ that produces $K_2O(s)$, $Na_2O(s)$, and additional nitrogen. Write a balanced equation for this process, showing the states of the reactants and products.

2) In this laboratory experiment you are to generate sufficient carbon dioxide to inflate a plastic zipper type bag. You first must carry out some calculations to determine the quantity of reagents needed.

The carbon dioxide will be generated by reacting solid sodium bicarbonate ($NaHCO_3$) with a solution of acetic acid (CH_3COOH). This is actually a two-step process that proceeds as follows:

Step 1: $NaHCO_3(s) + CH_3COOH(aq) \rightarrow \mathbf{H_2CO_3(aq)} + CH_3COO^-(aq) + Na^+(aq)$
Step 2: $\mathbf{H_2CO_3(aq)} \rightarrow H_2O(l) + CO_2(g)$

You measure the plastic bag dimensions to be 7.15 in. by 7.97 in. Assuming the average thickness of the bag when inflated will be 1.50 in. what do you calculate the volume of the bag to be?

Of course, this volume is only approximate, as the width of the inflated bag is not uniform. To get a better value for the volume, you fill the bag with water and then measure the volume of the water to be 1274.7 mL. Now, you measure room temperature to be 21 °C, and the barometer on the wall gives a reading of 756 torr for the atmospheric pressure. Using the ideal gas law, calculate the number of moles of CO_2 that will be required to fill the bag at this temperature and pressure.

Now, inspect the reactions above that are generating CO_2 and calculate the molar quantity of $NaHCO_3$ and CH_3COOH needed.

Quantity of $NaHCO_3$:

Quantity of CH_3COOH:

What mass of $NaHCO_3$ will be needed?

You have a solution of 6.03 M acetic acid. What volume of this solution will be needed?

Experiment 16
Intermolecular Forces and the Triple Point of CO_2

Introduction

The ability of matter to change from one state to another is vital in geological, chemical, and even physical realms. As we all know, water can naturally be found as a solid, liquid, and gas. What you may not know is just how matter changes from one form to another and the factors affecting such a process. This experiment is designed to further elucidate the concept of phase transitions using a rather elementary scientific approach. After completing this lab—with only a few chemicals, a thermometer, and a "home-made" barometer—you will better understand phase changes and the effects of temperature, pressure, and intermolecular forces on those changes.

Whether you are aware of it or not, every one of us has witnessed a change of phase, whether it was through the evaporation of sweat from our bodies while working out or something as simple as making a tray of ice cubes. Phase changes are important geologically, as there is growing concern for the melting of the polar ice caps caused by climate change. However, before we save the world from melting polar ice caps, we have to understand phase transitions.

We all know that solid water can easily be converted into its liquid form simply by adding heat, but the processes occurring at the molecular level are not as obvious. Recall that phase changes are physical, not chemical, changes. This means that instead of breaking intramolecular bonds—the bonds between atoms, the bonds between molecules are being broken. When the intermolecular attractions between water molecules in a drop of water are weakened and ultimately broken, water molecules escape into the gas phase. To emphasize the distinction between these intermolecular bonds and chemical bonds, we will instead refer to intermolecular attractions. To investigate this notion, we are going to observe the change in temperature that occurs in the surroundings when five different liquids are evaporated at room temperature. Then we will use this information to draw conclusions about the strength of each liquid's intermolecular forces.

The second part of this experiment will use a barometer to investigate the effect of temperature and pressure upon phase changes. A barometer can be any instrument that is used to measure atmospheric pressure. Temperature and pressure play an intricate role in the transition of matter from one phase to another, as shown by a phase diagram. However, graphs do not necessarily show how "explosive" such relationships can be, such as the one you will be investigating in this portion of the experiment.

The first part of this experiment will advance your knowledge of phases, the changes they may undergo (i.e., solid, liquid, or gas states), the underlying mechanisms responsible, and the effect of intermolecular forces upon them, and the second portion will show you the relationship between temperature and pressure in determining the triple point of CO_2.

Background

All matter occurs naturally in one of three physical states: solid, liquid, or gas. A fourth state, plasma, will not be considered here. (It requires significant magnetic fields for containment.) The term **state** of matter is used interchangeably with the term **phase**. A phase is defined as any physically distinct homogeneous part of a system. The phase or state of a substance is determined by both the energy of the system and by the intermolecular attractions between the molecules or atoms of which the substance is composed. According to the *kinetic molecular theory*, gases are composed of high energy molecules that do not experience intermolecular forces due to both the speed at which they travel and the space that exists between their molecules. On the other hand, the molecules of a liquid are much closer together and move much more slowly. Therefore, a number of intermolecular forces, ranging from the weak London dispersion forces to the much stronger networks of hydrogen bonds, are characteristic of liquids. Solids are composed of low energy molecules that are packed very closely together. The molecules in solids are said to vibrate, not move. Most ionic compounds are solids at room temperature because the ionic interactions they experience are extremely strong. The energy required to melt these ionic solids is normally very high, and the energy required to promote them into the gas phase is even greater. Ionic interactions are the strongest intermolecular forces.

Phase Changes

So how do phase changes take place? For a solid to change into a liquid (melt) or a liquid to change into a gas (evaporate), the intermolecular forces between the particles must be overcome. This requires the input of energy. The most common way to put energy into a system is by adding heat. You have experience with solid ice melting into liquid water and liquid water boiling to form gaseous steam, both phase changes resulting from the input of heat energy.

In this experiment, you will investigate the strength of the intermolecular forces of five liquids. Their identities are methanol, isopropanol, n-hexane, acetone, and water. The liquids will be provided to you as unknowns in bottles marked A through E. It is your job to identify each of the unknown liquids by determining the relative strength of their intermolecular forces and by making any other scientific observations you can. By comparing your observations to the known properties you gathered in your prelaboratory exercises, you should be able to identify the compounds.

Each of the five liquids will evaporate at room temperature. Those with stronger intermolecular forces will require more energy input to evaporate than those with weaker intermolecular forces because they have a higher enthalpy of vaporization (ΔH_{vap}). The enthalpy of vaporization is a physical quantity that measures the amount of energy required to vaporize the liquid, that is, cause the liquid to pass through a phase change from liquid to gas. Evaporation is an endothermic process that results in a drop in temperature of the surroundings. The compound requiring the least energy for vaporization (smallest ΔH_{vap}); thus, the weakest intermolecular forces will exhibit the largest temperature decrease. This occurs because as the molecules evaporate, they carry heat away from the thermometer which registers each loss as a drop in temperature. Molecules with stronger intermolecular forces require a larger amount of energy to accumulate before they can evaporate, so their removal of heat from the thermometer is a much slower process. The length of time between evaporation events allows the thermometer to recover heat from the atmosphere; thus, we do not observe as great a drop in temperature for these molecules. If we monitor this temperature change, we can relate the magnitude of the temperature drop to the number and relative strength of the intermolecular forces of each substance. Relating this information to their known ΔH_{vap} values allows us to determine the identity of each of the unknowns.

Phase Diagrams and the Triple Point of CO_2

The second part of this experiment investigates the triple point. Phase diagrams illustrate the temperatures and pressures at which matter changes from one phase to another. The triple point of a substance is the temperature and pressure at which all three states of matter exist simultaneously. For this experiment, we will determine the triple point of carbon dioxide. In its solid state, CO_2 is commonly referred to as dry ice. If you look at the phase diagram of CO_2, you can see that at room temperature (25 °C) and pressure (1 atm), CO_2 is a gas. What is also clear is that if dry ice (solid CO_2) is present at room temperature and pressure, it will not melt instead, it will sublime, that is, it will transition directly from solid to gaseous state. For us to be able to observe the liquid state of CO_2, the pressure must be at or above the pressure of the triple point (5.1 atm). In this experiment, the pressure of the CO_2 is raised by placing dry ice in a sealed pipette and allowing it to sublime at room temperature. As more and more gas forms, the pressure inside the pipette increases, hopefully to the triple point. You will use a Boyle's law-type microgauge to determine the pressure at the observed triple point (the point at which you observed the dry ice liquefy).

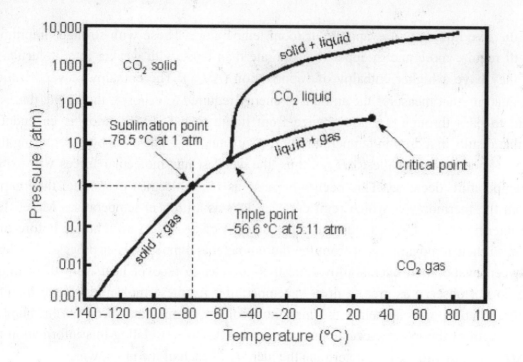

Pressure-Temperature phase diagram for CO_2.

Boyle's Law

Boyle's law: $P_1V_1 = P_2V_2$

Where P_1 is read from the barometer in the lab.

V_1 is the total volume of your microgauge.

V_2 is the final volume read from your microgauge at the triple point.

P_2 is the experimentally determined pressure at the observed triple point.

Boyle's law relates the volume of a system to its pressure. The microgauge we will construct will be made of a thin-nose plastic pipette sealed at one end. Normally, we would need to calculate the true volume of the pipet by measuring the radius of the tube and using the equation for the volume of a cylinder ($V_{cyl} = \pi r^2 h$) to determine its volume. However, the pipette's diameter will remain constant throughout the experiment, and only the length portion (h) of the volume equation will change, so we can omit the volume calculation on the principle that if a value is equal on two sides of an equation, that value will cancel itself out. If we then complete the experiment and substitute the values into Boyle's law in the manner described above, we should be able to measure the approximate triple point of CO_2. I believe you will be surprised at how accurate your crude microgauge can be.

Procedure

SAFETY NOTES:

Part I: All of the organic solvents used in this part of the experiment are flammable. No flames should be used in lab today. The fumes of some of the solvents can be irritating. Please use your hand to waft the solutions toward you to smell them if necessary. Do not rinse any of the solvents down the sink. There are waste containers provided in the hood. Avoid direct contact with the solvents, as they may cause irritation. If you should get any of the solvents on your skin, wash the exposed skin with soap and water as soon as possible and notify your instructor.

Part II: Dry ice can cause frostbite if held in the hand too long. Minimize direct contact with the dry ice as much as possible. Make sure the pipette bulb is submerged in the water while the dry ice is subliming. There is always the possibility that the bulb will burst under the pressure being exerted. The water in the cup will absorb the concussion and the debris. (NOTE: You might get wet from the splash.)

Part I: Evaporation and Intermolecular Forces

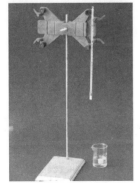

Collect a thermometer.

Prepare the thermometer by wrapping small cut pieces of filter paper secured by small rubber bands around the tip. Roll the filter paper around the tip in the shape of a cylinder and then slip the rubber band(s) around the filter paper to secure it in place. Make sure that the filter paper is even with the end of the thermometer.

Collect a stand and clamp from the front counter. Place a clamp adaptor around the thermometer and clamp it in place. Make sure you can read the thermometer from ~30 °C and below.

Place ~10 mL of one of the unknown liquids A through E in a 100 mL beaker.

Lift the beaker of unknown liquid under the thermometer, submerge the tip in the liquid, and hold it there for at least 30 sec.

Begin collecting temperature data while the tip is still submerged. Take several readings for at least 1 min; the average of these readings will be your T_i.

Lower the liquid away from the thermometer and observe the temperature change that occurs. Continue to take temperature readings until the temperatures pass through a minimum, this will be your T_f.

When done, clean the probe with DI water and repeat the above process for each of the unknown liquid, using a new piece of filter paper each time.

Part II: The Triple Point of CO_2

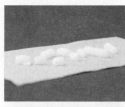

Collect some dry ice (you need only a few small pieces).

Place the dry ice on a paper towel on the lab bench and watch what happens. Take notes.

Take a plastic pipette and cut off the graduated tip.

Crush some dry ice inside a folded paper towel.

Place enough dry ice "crumbs" into the pipette to fill the bulb up about one-third full.

Fill a clear plastic cup to a depth of about 7 to 8 cm with water.
Fold the tip of the pipette bulb shut and clamp tightly using the pliers provided. Make sure the seal is very tight so that no gas can escape.

Using the pliers, immediately lower the bulb portion of the pipette into the water.
Observe from the side of the cup what happens. Take notes. You can also release and tighten the pliers. Observe what occurs with the CO_2 each time you do.

Collect a plastic thin-stem pipette. Using the technique demonstrated by your lab instructor or teaching/lab assistant, cut the "nose" off of the pipette.

Collect a piece of string and use the hot glue gun provided to attach the string to the pipette nose. Be sure that the glue completely seals one end of the pipette nose.

Using a wax pencil and a ruler, mark the newly created pipette microgauge every centimeter from the sealed end as far up as you can. "0" is maked at the point where the glue seals the tube.

Place a drop of food coloring on a watch glass and add a few drops of water to dilute it. (Everyone at a bench can share.)
Cut the end of the microgauge at an angle and add a drop of food coloring to the open end of the microgauge.

Fill another plastic pipette with dry ice as before. Insert the microgauge into the pipette bulb open end down but suspended by the string such that the microgauge is just above the CO_2.

Clamp the pipette shut, making sure you don't crush the microgauge, just the string.

 Place the pipette with the microgauge into the cup of water as you did earlier. Pay close attention to the microgauge and record the change in volume at the observed triple point (the point at which you first observe liquid CO_2) by determining the distance the food coloring has traveled at that point. Read the microgauge from the bottom of the food coloring mark.

Report Contents and Questions

The purpose should be comprehensive for both Part I and Part II. This section should include the various topics covered throughout the experiment. The procedure section must cite the lab manual, noting any changes made to the written procedure.

For Part I, you should report the ΔT values for each of your unknown liquids. Based on those ΔT values, rank the intermolecular forces for unknown liquids A through E. Using the rankings and what you know about the molecular structure of the liquids, propose identities of the unknown liquids. Explain your choices with all of the supporting evidence you collected.

For Part II, report all of your observations of the CO_2. Use Boyle's law to calculate the pressure at the triple point determined by your microgauge.

Once the value of P_2 is known, the percent error of your microgauge can be determined. Before calculating percent error, make sure the units for both triple points are the same. The conclusion section should be several paragraphs that address the following: For Part I, the identity of the five unknown liquids with an explanation supporting the identification made and a discussion of the similarities and differences of your rankings compared to the heat of vaporization ranking done in the prelab activities. For Part II, summarize the CO_2 observations; state and compare the actual and experimental values for the triplet point of CO_2, including the percent error. Finally, discuss any errors in the experiment

Answer the following questions:

1) How is the change in temperature we observe as the various solvents evaporate related to their ΔH_{vap}? Explain this relationship.

2) In our use of the microgauge to determine the triple point pressure of CO_2, we measure only the height change (distance the dye moves) in the gauge, not the actual volume. Why does this approximation not ruin our results?

Experiment 16
Laboratory Preparation

Name: _____ Date: _____

Instructor: _____ Sec. #: _____

Show all work for full credit.

1) Read the background, procedure, and report sections of the lab experiment carefully and develop a hypothesis of what information you expect to gain from the completion of the lab experiment.

2) Create any and all tables you might need to collect data during the experiment and then transfer those tables into your lab manual for use during the lab.

Experiment 16
Prelaboratory Assignment

Name: _____ Date: _____

Instructor: _____ Sec. #: _____

Show all work for full credit.

Copy the information gathered in Parts I and II of this prelab into your lab notebook for later use.

1) For the following liquids, methanol, isopropanol, n-hexane, acetone, and water, (a) write the molecular formula, (b) draw a Lewis structure for each compound, and (c) calculate the molecular weight of each compound.

2) The relative strength of the intermolecular forces depends on the size (i.e., MW) and structure of the molecules and include, in rough order of decreasing strength, ion–ion, ion–dipole, hydrogen bonding, dipole–dipole (polar), dipole–induced dipole, and induced dipole–induced dipole (dispersion) forces. The latter increases with molecular weight and can sometimes become the most predominant force. There are no ions in these liquids, and all the molecules are alike, leaving primarily hydrogen bonding, dipole–dipole, and disperson forces to consider.

What are the strongest intermolecular forces in acetone?
What are the strongest intermolecular forces in methanol?
What are the strongest intermolecular forces in 2-propanol?
What are the strongest intermolecular forces in water?
What are the strongest intermolecular forces in hexane?

3) The figure at the right shows your experimental setup for estimating the triple point of CO_2. Carefully mark off your capillary tube in ten 1 cm segments and note the position of the drop of dye solution when the capillary is inserted in the open plastic tube. When the top of the plastic tube is sealed, the pressure inside will increase as the CO_2 sublimes, and the drop of dye will be forced up the capillary, compressing the air inside it. Note the position of the drop of dye when liquid CO_2 starts to form. The pressure inside the capillary should then correspond to the pressure at the triple point of CO_2.

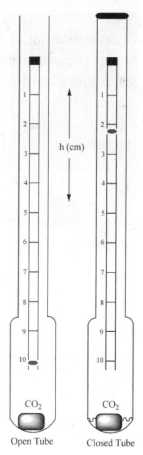

Open Tube   Closed Tube

You can calculate the pressure from a modification of Boyle's law, $P_1V_1 = P_2V_2$. The volume of the capillary = hA, where h is the height and A is the cross-sectional area, which is constant. Boyle's law therefore reduces to $P_1h_1 = P_2h_2$, where P_1 is the pressure in the open tube, h_1 is the distance of the drop from the top of the capillary in the open tube, h_2 is the distance of the drop from the top of the capillary in the closed tube, and P_2 is the pressure in the closed tube, the pressure at the triple point.

What are the heights of the air columns in the two capillaries to the right? (Measure from the top of the dye droplet.)

Height of column in the open tube (h_1):
Height of column in the closed tube (h_2):

Calculate the pressure in the closed tube, assuming the pressure in the open tube is 1.0 atm.

Pressure in the closed tube:

Experiment 17
The Purification of Water

Introduction

Clean drinking water is of great importance to people worldwide. The purification of drinking water is a laborious and complicated process. In this laboratory, you are introduced to the various processes by which water quality is assessed and treated. As you are guided through these processes, you will also be introduced to a number of valuable new chemical techniques.

Nothing is more vital to the survival of living organisms than water. In fact, ~60% of the human body is water; the brain is composed of ~70% water, and the lungs are nearly ~90% water. Approximately 83% of our blood is water, which helps digest our food, transport waste, and control body temperature. It is estimated that each day, humans must have about 2.5 L of drinking water to maintain a reasonable quality of life. As the earth's population continues to grow, an adequate supply of clean water is becoming a challenge. Water may be everywhere one looks and cover approximately two-thirds of the world, but only a small amount of that water is considered drinkable.

Biological or chemical impurities can make water unfit for drinking. Bacteria, algae, viruses, fungi, minerals, and other chemicals can contaminate water sources such as rivers, lakes, and oceans. Some of these simply make water unappealing in smell, taste, or appearance; others are dangerous. Most of these impurities must be removed for the water to be considered drinkable.

Background

The first step in processing any water is the collection of samples. As with any physical sample collection, the location and specific physical parameters (temperature, pH, sediment content etc.) will affect the overall appearance and chemistry of your sample. The procedure provided here for the proper collection of your water sample is designed to establish good collection techniques for your future as a scientist. Why is this so important? Suppose you collect your water sample in an old bleach bottle. If the bottle still contains remnants of the bleach, it will contaminate your water sample. Another example would be if you collected your water sample just after a rainstorm but did not note this in your report. It is likely that the levels of contamination in your samples would be considerably lower than in samples taken prior to the rain, as the rain would have diluted the contaminants.

The second step in this experiment is the analysis of your collected water sample for possible contamination. This establishes a baseline from which you can determine the effectiveness of your purification process. Most municipalities have entire departments dedicated to monitoring water quality in their lakes, ponds, and drinking water. The process we will use to determine the phosphate content of water samples is identical to the one used across the country by many of these labs. Specifically, we will use an analytical technique called absorption spectrophotometry. Details of the specific type of spectrophotometry used in this experiment are outlined in the background section. In addition to sample collection and absorption spectrophotometry, this experiment introduces you to several other new techniques that are invaluable to a chemist: serial dilution, filtration, and pH.

Spectrophotometry is the study of the interaction of electromagnetic radiation with matter. Through the many spectroscopic techniques available, the shape, the composition, and the way a compound or molecule reacts can be determined. It is a widely used technique with several variations, including Raman, UV-visible, infrared, and nuclear magnetic resonance. Each of these types of spectrophotometry assesses the interaction between the sample and a different type of radiation. The actual instrumentation and the process of preparing samples for each of these techniques are unique, but the general concept of spectrophotometry remains the same. A basic understanding of how to use a spectrometer to determine the concentration of a sample is yet another skill you should acquire as a chemist.

Serial dilution is a process that is vital to the conservation of chemical resources in a lab. For example, if you need approximately 10 mL each of a series of solutions from 1 M to 5 M spaced at half-molar intervals (e.g. 1 M, 1.5 M, 2 M, 2.5 M…), you could simply use a 5 M solution as stock and take enough stock solution to make each of the more dilute solutions. Making all these solutions would take a very long time and use a large amount of stock solution. Serial dilution, however, would allow you to make these mixtures using a relatively small amount of stock solution. In a true serial dilution, each solution becomes the stock solution for the next solution needed, allowing you to work very efficiently.

In this lab, we will create colored solutions of known concentration using serial dilutions and then use a spectrophotometer to determine their absorption value. This will allow us to produce a calibration curve by which we can determine the concentration of the water samples collected in the field. The use of graphical analysis to determine an unknown value is another recurring theme in the lab and a handy technique you should become competent in performing.

Other techniques are explored during the purification process. Precipitation followed by filtration is a process by which some contaminants can be separated from a solution. Vacuum filtration uses suction to increase the speed at which you can filter your water. Precipitation uses chemical reactions to remove unwanted ions from your water. Referencing the solubility tables in your course textbook, you should find that the addition of CaO and $Al_2(SO_4)_3$ produces precipitates that result in the removal of various ions from your water samples. Filtrations (both gravity and vacuum) are then used to separate those precipitates from your water samples.

Another technique we will introduce is the use of a pH meter to determine the acidity or basicity of the water sample. pH determination can indicate whether our water sample is contaminated and if it is, provide clues as to the nature of the contaminant. Before you use a pH meter, it must first be calibrated to make sure that it is reading accurately. Your instructor will guide you through the calibration of the pH meters used in your lab.

As a conclusion to the lab, you are asked to repeat the phosphate determination, this time using your "purified" water sample. This assessment determines the success of the process you used to purify your samples.

Water Contamination

Chemical impurities found in water include many dissolved salts. Ions such as potassium, calcium, magnesium, iron, sulfate, and carbonate become dissolved in water as rain and groundwater pass over rocks, which slowly dissolve into the water. Chemical impurities can also come from human sources. Two common chemical impurities are nitrite (NO_2^-) and phosphate (PO_4^{3-}). Nitrite contamination is caused by fertilizer runoff from farms and agricultural lands, while phosphate contamination is caused by fertilizers, wastewater treatment plants, soaps, detergents, and industrial processes.

Being able to test for the presence of both biological and chemical impurities in water is important. This experiment is designed to give you experience in some of the testing and purification processes commonly performed in a water quality lab or water treatment plant. BEFORE your lab period, you MUST collect approximately a liter of water from a natural water source (e.g., a lake, river, or runoff in the area. Once in lab, the following techniques will be used to test for the concentration of phosphate in the water sample and purify the water: a test of pH, spectroscopic determination of the concentration of phosphate, serial dilutions to prepare standards, and filtration to remove sediment and other solids.

pH

pH is a good indicator of water contamination. The pH of a neutral solution is 7, while an acidic solution is less than 7 and a basic solution is greater than 7. The pH of natural source water is going to vary because of several factors. The pH may be basic because of photosynthesis. Photosynthesis uses dissolved CO_2, which acts like carbonic acid in water. When water plants use the dissolved CO_2, the acidity is reduced. The presence of large quantities of bacteria or other biological contaminants in water can produce ammonia that will raise the pH of the water. Acidic water can result from acid rain or from the presence of much industrial waste, including sulfur and nitrogen oxides which become sulfuric and nitric acids, respectively, when carried by rain or runoff into lakes or streams.

The pH is normally recorded in the field at the water source, but as long as the water sample is refrigerated to keep any biological entities from changing the pH, it should be unchanged at the time of the experiment. For this experiment, the pH will be measured and recorded in your lab notebook using a pH meter. The pH will be measured twice, before and after the purification process.

Determining Phosphate Concentration

The phosphate concentration will be measured using spectrophotometry. To understand how spectrophotometry works, one must first understand the basic principles of light and energy. When an electron absorbs energy, that is, light, the electron moves from one energy level to the next, usually from the ground state to an excited state. An absorption spectrum is produced when continuous electromagnetic radiation (every wavelength, all colors) passes through a substance and certain wavelengths are absorbed. In other words, an observed absorption spectrum is made up of the wavelengths of light that are not being absorbed by the substance. A molecule can absorb light only if it can accommodate the additional energy by promoting electrons to higher energy levels. The energy of the light being absorbed must match the energy required to promote an electron. Therefore, not all wavelengths of light are absorbed equally by a sample.

In this experiment, we will be using a UV-Vis spectrometer to determine the concentration of phosphate in your water sample. A color–treated water sample is placed into a cuvette and then placed into the instrument. Inside the spectrometer, light is passed through the sample cell at the appropriate wavelength selected by a diffraction grating. The light can scatter, be absorbed by the molecule of interest and re-emitted in any direction, or pass through the sample cell without interacting with the sample at all. Because it is concentration dependent, we are interested in measuring the amount of light absorbed and emitted. The amount of light that passes through the sample is quantified by the detector.

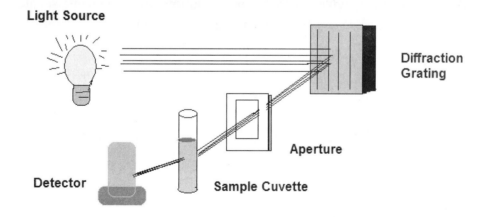

Spectrometers like the one above can be set to read in two different units, either absorbance (A) or percent transmission (%T). In this experiment, we will set the spectrometer to read absorbance (A) because the Beer-Lambert law expresses the direct correlation between the absorbance of a sample, its concentration, and its thickness. The law can be written as:

$$A = \varepsilon bc$$

where ε is the molar absorptivity (this is a constant that depends on the nature of the absorbing system and the wavelength passing through), b is the path length (the width of the sample cell or cuvette, usually 1 cm), c is the concentration of the sample, and A is the absorbance of the sample.

If the molar absorptivity and the path length are held constant, the relationship between absorbance and concentration can be easily studied. If there is a direct relationship, a plot of different concentrations of the solution will generate a straight line, and the system is said to obey the Beer-Lambert law. When a curved plot is obtained, the system deviates from the law, and the correlation between concentration and absorbance cannot be easily established. (This can happen at high concentrations.)

Colorimetric Methods

Since the phosphate ion in water is colorless and therefore not easily detected using spectrophotometric techniques (which determine concentration from color intensity), the phosphate must be reacted with a color-producing agent. In this experiment, we will react the phosphate with ammonium molybdate $((NH_4)_6Mo_7O_{24} \cdot 4H_2O)$ and stannous chloride $(SnCl_2 \cdot 2H_2O)$ in an acidic solution. When combined with phosphate, these reagents produce a blue compound whose absorption can be observed at 650 nm. Measuring the absorbance of a set of standards with known concentrations of phosphate will allow us to develop a linear graph that can be used to determine concentrations of unknowns based on their absorbance values. The absorbance value of your water sample can then be used to determine the concentration of phosphate in your water.

Serial Dilution

Serial dilutions are a way to make solutions of varying concentrations very easily in succession. In this experiment, you will be provided with a stock solution of 10 ppm phosphate. From that solution, you will make phosphate solutions of 8 ppm, 6 ppm, 4 ppm, and 2 ppm. It is extremely important to understand how to make serial dilutions before coming to lab.

Example:

Using a 15 ppm stock solution, create solutions of 10 ppm, 7 ppm, and 3 ppm using serial dilution. Sample cuvettes only need about 2 to 3 mL for a measurement so starting with 10.0 mL of 15 ppm stock and using the equation

$$M_1V_1 = M_2V_2$$

where M_1 is concentration of solution one, V_1 is the volume of solution one, M_2 is concentration of solution two, and V_2 is volume of solution two, you can calculate the amount of stock solution you will need to make 10.0 mL of 10 ppm solution:

$$(10.0 \text{ mL})(10 \text{ ppm}) = X(15 \text{ ppm})$$

$$X = 6.7 \text{ mL}$$

The results indicate that 6.7 mL of 15 ppm solution contains the appropriate amount of phosphate to make a 10 ppm solution when diluted to 10.0 mL. The 10 ppm solution is therefore made by measuring out 6.7 mL of the 15 ppm stock phosphate solution and then adding 3.3 mL of water to make a 10.0 mL total volume.

This process is repeated to make the 7 ppm solution. But the stock solution used this time is the newly made 10 ppm solution. Repeating the calculations as above:

$$(10.0 \text{ mL})(7 \text{ ppm}) = X(10 \text{ ppm})$$

$$X = 7 \text{ mL}$$

The 7 ppm solution is therefore made by measuring out 7.0 mL of the 10 ppm phosphate solution and then adding 3.0 mL of water to make a 10.0 mL total volume. The process continues, using each newly made solution as the stock for the next lower concentration until all of the needed concentrations are produced. The results are a series of solutions that are similar in appearance to the ones below (color intensity decreases as concentration decreases):

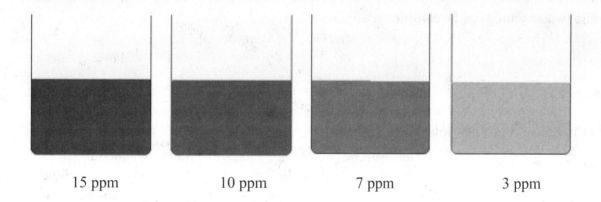

| 15 ppm | 10 ppm | 7 ppm | 3 ppm |

Phosphate Concentration

After the solutions are prepared by serial dilution, the absorbance of each sample must be determined using the spectrometer. Looking at the Beer-Lambert law, you can see that if you plot absorbance versus concentration, the slope of the line is the path length multiplied by the molar absorptivity (εb).

$$A = \varepsilon bc$$

$$y = mx$$

Since the value of εb is a constant, the concentration of phosphate in the water sample you collected can be determined from the absorbance of the sample. The absorbance value is measured in the same manner as absorbances were obtained for the standards. The Beer-Lambert law is then rearranged and solved for the sample's concentration (c). For example, if the absorbance is 0.253 a.u. and the value of εb is 1.45×10^4 M, you can solve for the concentration of the sample:

$$A = \varepsilon bc \Rightarrow$$
$$0.253 = (1.45 \times 10^4 \; \text{L}/_{\text{mol} \cdot \text{cm}})(1 \; \text{cm})(c) \Rightarrow$$
$$c = 1.74 \times 10^{-5} \text{M}$$

Water Purification

To make the natural water sample you collected drinkable, you must remove the chemical and biological impurities. The first step is to filter the water through filter paper to remove large particles from your sample. The next few steps of the process remove the remaining ionic impurities through the addition of chemical reagents and several more filtrations. Once the filtration steps are complete, the pH and absorbance measurements are repeated to determine if the natural water source can be considered drinkable.

Procedure

Part I: Collection of Samples

You will be collecting your own water sample for use in this laboratory. *The collection should take place no more than 24 hr before lab.* This is because analysis must take place within 48 hr to be valid. The sample should come from a fresh water source because specialized tests that we are not equipped to run required to analyze salt water samples.

Obtain a clean and dry ~1 L glass or plastic bottle. You may use a 20 oz + or larger soda bottle if you clean it thoroughly with hot water and rinse it with vinegar. Do not use soap, as doing so might add phosphate.

Find a source of fresh water. Take copious notes regarding the location and condition of the source.
Collect the sample by inserting the bottle upside down into the water. Rotate the open end toward the direction of the flow and allow the bottle to fill under the surface. Sample about 12 in. under the surface. Avoid surface scum and bottom sediment.

Label the bottle with the sample name, location and time collected, and the initials of who collected the sample.

Store it in a refrigerator until it is time for lab. (You can bring samples by the lab early in the day or keep them on ice in a small cooler until lab.)

Part II: Analysis of pH

Take 50 mL of your water sample and place it in a 100 mL beaker that has been cleaned and rinsed with the acid wash and distilled water provided.

Calibrate the pH meter as instructed by your instructor or TA.

Using the pH meter, take the pH of the sample and record it in your notebook.

Part III: Analysis of PO_4^{3-}

Turn on your spectrometer.

Take 25.0 mL of your water sample, 25.0 mL of stock phosphate solution, and 25.0 mL of distilled water. Place each of them in separate 100 mL beakers that have been cleaned and rinsed with the acid wash provided.

To the beakers containing the unknown water samples (one from each student at the bench) and to the phosphate standard beaker (only one, which is shared) add 1 mL of the Ammonium Molybdate color reagent and two drops of stannous chloride.

Mix the solutions for 5 min allowing color to develop.

Using the calculations from your prelab activities, create the four serial dilutions of your stock solution. These will be used to create the standard calibration graph to determine the concentration of your unknown, so be meticulous in your technique when making the dilutions.

You are now ready to measure the absorbance of your samples.
Make sure your spectrometer has been on for a minimum of 10 min. Obtain a cuvette from the instructor or teaching/lab assistant. Do not break it; be sure to return it at the end of the class.

Wash the cuvette carefully and rinse it with distilled water. Do not use a brush to wash the cuvette, as they scratch very easily.
Fill the cuvette to about two-thirds full with your distilled water blank.

Press the A/T/C button on the spectrometer to select the absorbance mode. The current mode appears on the display.

Using the nm (up/down) controls, set the wavelength at 650 nm.

Observe the white reference mark on the cuvette. Align this mark with the one on the sample compartment every time a measurement is taken.
Insert the blank cuvette into the sample compartment and close the compartment's lid.

Press 0 ABS/100%T to set the blank to 0 absorbance. Remove the blank cuvette and discard the blank water sample.

Rinse the sample cuvette with a small portion of the standard stock solution (2.50 g/L) and discard the rinse. Wipe any fingerprints off the outside of the cuvette with a Kimwipe and handle the cuvette with a Kimwipe from here on. Fill the cuvette about two-thirds full with the stock solution and insert it into the sample compartment. Record the absorbance.

Remove the sample cuvette, discard the stock solution in the waste jar, and rinse the cuvette well with DI water.

Rinse the sample cuvette again with a small portion of your next standard solution and discard the rinse. Wipe any fingerprints off the outside of the cuvette with a Kimwipe and handle the cuvette with a Kimwipe from here on. Fill the cuvette about two-thirds full with the standard solution and insert it into the sample compartment. Record the absorbance.

Repeat until all the standards and all of the student's unknown samples have been measured.

Part IV: Water Purification (student should purify their own water sample)

Using the 100 mL of your water samples previously used for measuring pH. *Slowly* filter the sample using a suction filtration setup to remove large particulates. Use 0.1 mm filter paper.

After a liquid sample has been passed through a filtration system, it is called a filtrate.

Add one spatula of dry lime (CaO) powder and one spatula of dry alum (Al$_2$SO$_4$) powder to the filtrate and stir with a clean glass stirring rod. Allow the powders to react for ~10 min. Make observations.

Using your filter funnel again, place a filter paper in the funnel and then add enough sand to cover the filter paper to a depth of ~0.5 cm with sand.
Slowly pass the filtrate through the sand.

Add 1 mL of sodium hypochlorite (NaHClO) reagent to the filtrate. Stir to mix completely.

Set up a charcoal filtration funnel by placing a filter paper in the Buchner funnel. Then add a 1 cm layer of activated charcoal on the filter paper and cover with a second filter paper. Set up a vacuum filtration as instructed. Use an adaptor and fit the funnel into the flask.
Rinse the charcoal with distilled water until the water in the flask is clear. Make sure you discard the charcoal rinse water and dry the filtration flask before filtering your water sample.
Pass your water sample through the fully rinsed charcoal filter.
Make observations regarding the color and clarity of the "purified" water.

Part V: Water Quality after Purification

Repeat the pH test on your purified water sample. If the pH is >8, add a drop of 1.0 M HCl to neutralize it.

Repeat the phosphate test for your purified water sample only. You don't need to repeat the measurement for the standards.

Report Contents and Questions

The purpose should be several sentences summarizing the objectives of the experiment and the criteria for determining whether those objectives were achieved.

The procedure section should reference the lab manual and note any changes made to the experiment. For this experiment, the procedure must include the location, date, and time of sample collection. Also be sure to note any changes made to the procedure and the names of your team members.

The data section should include a data table with the phosphate concentrations and absorbance values of the five standard solutions and your collected water sample before and after purification. The concentration of phosphate in your water sample before and after purification should be reported BOTH in ppm and molarity. [1 ppm = 1 mg/L; 1 mole PO_4^{-3} = 9.5 × 10^4 mg PO_4^{-3}] The pH of the water sample before and after purification must also be reported. Include all observations of your water sample made before, during and after purification.

For the calculation section prepare a graph of concentration vs. absorbance for PO_4^{-3}. All graphing guidelines apply here. Also include sample calculations of the concentration of the phosphate concentration of your water sample before and after purification using Beer's Law ($A = \varepsilon bc$, where A is the absorbance; ε is the molar absorptivity, which is equal to the slope of the line; b is 1.00 cm; and c is the concentration). A sample calculation of the conversion from ppm to molarity should also be included. Finally, you must include an example calculation of the purity of the water sample. This can be determined using the following formula:

$$\%purity = \frac{ppm\ PO_4^{-3}\ before\ purification\ -\ ppm\ PO_4^{-3}\ after\ purification}{ppm\ PO_4^{-3}\ before\ purification} \times 100\%$$

The conclusion should be several paragraphs addressing the following: The location where the water sample was obtained; the phosphate concentration and pH of the water sample before and after purification; a discussion of the possible reasons for the phosphate concentration level and for the pH value of the water sample before and after; in other words what might be causing these values. Discuss the purity of your water sample before treatment. As always, discuss any possible errors in the experiment.

Answer the following questions:

1) Why should we care about the amount of PO_4^{-3} found in a natural water source?

2) If you really wanted to be sure your water was drinkable, what further purification step(s) would you take?

3) Blue light has a wavelength between 450 and 495 nm. Why don't we measure the absorbance of a blue solution within this wavelength range?

Experiment 17
Laboratory Preparation

Name: _____ Date: _____

Instructor: _____ Sec. #: _____

Show all work for full credit.

1) Read the background, procedure, and report sections of the lab experiment carefully and develop a hypothesis of what information you expect to gain from the completion of the lab experiment.

2) Create any and all tables you might need to collect data during the experiment and then transfer those tables into your lab manual for use during the lab.

Experiment 17
Prelaboratory Assignment

Name: _____ Date: _____

Instructor: _____ Sec. #: _____

Show all work for full credit.

1) Calculate the $[PO_4^{3-}]$ concentrations corresponding to the following ppm values:

Concentration = 0.37 ppm; $[PO_4^{3-}]$ =
Concentration = 333 ppm; $[PO_4^{3-}]$ =
Concentration = 15 ppm; $[PO_4^{3-}]$ =

Calculate the ppm values corresponding to the following $[PO_4^{3-}]$ concentrations:

$[PO_4^{3-}] = 7.90 \times 10^{-5}$ M; Concentration =
$[PO_4^{3-}] = 2.42 \times 10^{-6}$ M; Concentration =
$[PO_4^{3-}] = 6.75 \times 10^{-3}$ M; Concentration =

2) Use the following table of standard concentrations and absorbances for the phosphate ion to create a calibration graph. What are the slope and intercept of the calibration curve?

What is the concentration of phosphate in a solution that has an absorbance of 0.708 AU?

One of your samples shows a reading of 1.541 AU, which is beyond the range of the calibration curve. The lab is over, and it is too late to redo the calibration over a broader range or to dilute the sample and determine it a second time. Assume that Beer's law still holds up to the value of the sample and calculate the concentration using the equation for the calibration line.
(Path length = 1.0 cm)

Phosphate Analysis	
Concentration (ppm)	Absorbance (AU)
0	0.040
2	0.288
4	0.466
6	0.635
8	0.833
10	0.991

3) Phosphate standards used to generate the calibration curve are usually made by first preparing a stock solution of known concentration and then diluting it to the required concentration.

First, prepare a stock solution that is 1.00×10^{-2} M in phosphate ion. How many grams of Na_3PO_4 would you add to a 250 mL volumetric flask to prepare this solution?
Mass Na_3PO_4:
What is this concentration in ppm of phosphate? (That is, as PO_4^{3-}, not Na_3PO_4).
PO_4^{3-} concentration:

What volume of this ppm solution would you dilute with water into a 1.000 L volumetric flask to make a solution of 10.0 ppm?

This much of the work will be done for you by the laboratory assistant. You will be given a stock solution of 10.0 ppm phosphate to use in calibrating your phosphate assay. To preserve the ammonium molybdate reagent, you will create the calibrating standards by doing a serial dilution, using an aliquot of each diluted solution to create the next dilution.

After adding 1 mL of ammonium molybdate reagent and 2 drops of stannous chloride to 25 mL of the 10.0 ppm solution to develop the color, use a 20 mL aliquot to dilute with 5 mL of water to make 25 mL of a solution of 8.0 ppm.
Using the dilution equation: $V_1 \times Conc_1 = V_2 \times Conc_2$ and rearranging:

$$20 \text{ mL} = 25 \text{ mL} \frac{8 \text{ ppm}}{10 \text{ ppm}}$$

To create 25 mL of a 6.0 ppm solution, what volume of the 8.0 ppm solution and what volume of water should be mixed?

To create 25 mL of a 4.0 ppm solution, what volume of the 6.0 ppm solution and what volume of water should be mixed?

To create 25 mL of a 2.0 ppm solution, what volume of the 4.0 ppm solution and what volume of water should be mixed?

4) What is the molar concentration of each of your calibration solutions?

Concentration = 10 ppm; $[PO_4^{3-}]$ =
Concentration = 8 ppm; $[PO_4^{3-}]$ =
Concentration = 6 ppm; $[PO_4^{3-}]$ =
Concentration = 4 ppm; $[PO_4^{3-}]$ =
Concentration = 2 ppm; $[PO_4^{3-}]$ =

Experiment 18
Kinetics: The Iodine Clock

Introduction

Chemical kinetics is a topic that is conceptually difficult for a number of students. The rate laws that are created and studied are as varied as and even more complicated than the chemical reactions they represent. In the laboratory, we will explore the concept of reaction kinetics by actually measuring the rate at which a reaction takes place.

One of the questions that often remains unanswered in our desire to teach a student all of the complex mathematics that go along with kinetic calculations is why kinetics is an important topic. The most readily recognized use of kinetics is in manufacturing. In the production of various chemical, it is vital to know the rate at which a product is both produced and broken down.

For example, in the production of steel, the proportions of iron and carbon are varied based on the intended use of the steel. The cooling of the molten steel (known as austempering) is controlled to allow a desired crystal structure to develop; these crystal structures have specific names and properties. When bainite (a type of iron) forms in austenite (a mixture of iron and carbon), the carbon content of the austenite increases as the transformation occurs. If the austempering stage is too short, martensite may form in the residual austenite. If it is too long, carbide precipitation occurs. Both phenomena are detrimental to the mechanical properties of the steel that is being produced. It is therefore necessary to know how long the austempering treatment should be.

Another more commonplace example is that of expiration dates or shelf lives of foods and drugs. Knowledge of the kinetics of the reactions that break down certain chemicals present in a food or drug can be used to determine how potent or unsafe a food or drug may be after a certain period of time. The same knowledge of reaction rates for specific chemicals can also be used to determine the dosage of a drug. For example, if the drug is slow to react within the body, the dosage can be larger because it is naturally time-released and won't hit the system all at once. If the reaction rate is very fast, too high a dose might kill the patient. Determination of reaction rates is of vital importance to the creation of chemical products and to the metabolism of life-saving drugs as well.

Our purpose in this laboratory is to determine the rate of reaction for the "iodine clock." The iodine clock reaction is a good example to use because the rate law can be determined by observing the completion rate of the reaction based on colorimetric evidence. Although you will not be learning any new techniques in this lab, you will get the opportunity to practice what you have learned previously regarding the proper way to make solutions. Since this is a quantitative laboratory, good quantitative transfer technique is vital.

Background

Chemical kinetics is the study of reaction rates and reaction mechanisms. In studying the chemical kinetics of a system, we are interested in what the reaction rate means, how to determine the reaction rate experimentally, how temperature and concentration affect the reaction rate, and the detailed pathway taken by the atoms as the reaction proceeds.

Measuring Rates Experimentally

The fundamental observation in reaction kinetics is of the reactants being consumed or the products being produced in a given time interval. Concentration data may then be used to calculate the rate of reaction for a given experiment according to the following equations.

$$\text{rate of reaction} = \frac{\text{amount of reactant consumed}}{\text{time of observation}} = \frac{-d[\text{reactant}]}{dt}$$

or

$$\text{rate of reaction} = \frac{\text{amount of product created}}{\text{time of observation}} = \frac{d[\text{product}]}{dt}$$

To measure a reaction rate, there must be some method of determining the concentration of reactants or products at the beginning and at the end of a time period. For this experiment, we will vary the initial concentration of a reactant and measure the time until that reactant is completely consumed. For most reactions, the rate decreases continuously with the progress of the reaction. This is due to decreasing concentrations of kinetically significant reactants. For example, assume a reaction of the general type:

$$A + B \rightarrow 2C$$

The rate law would be expressed as follows:

$$\text{rate} = \frac{-d[A]}{dt} = \frac{-d[B]}{dt} = \frac{1}{2}\frac{d[C]}{dt} = k[A]^m[B]^n$$

Assuming the superscripts m and n are positive integers, the rate of the reaction will decrease as the number of moles of A and B diminish as they form C. The exponents m and n in the above equation are called reaction orders, and their sum is the overall reaction order. For example, if $m = n = 1$, the reaction order in A is 1, the reaction order in B is 1; the reaction is first order in A and B; and the overall reaction order is $1 + 1 = 2$, or second order. Both m and n are determined experimentally and cannot be deduced from the stoichiometry of the reaction.

$$\textbf{rate} = k[A]^1[B]^1 \qquad \text{overall reaction order is 2}$$

220

The Iodine Clock Reaction

In this experiment, we will investigate the kinetics of the following reaction, known as the iodine clock. The iodine clock reaction (or Landolt reaction) is a classic chemistry reaction in which two colorless solutions are mixed, and their chemical reaction results in a colored solution. This reaction demonstrates that reaction rates depend on the concentrations of the reagents involved in the overall reaction. In the iodine clock reaction, you are measuring the rate of the reaction of iodide ion (I^-) with peroxydisulfate ion ($S_2O_8^{2-}$):

$$S_2O_8^{2-} + 2I^- \rightarrow 2SO_4^{2-} + I_2 \quad \text{Reaction 1}$$

The rate of the reaction is determined by measuring the amount of I_2 formed over a particular amount time, or $\Delta[I_2]/\Delta t$. The I_2 is measured by titrating it with a fixed amount of thiosulfate ion ($S_2O_3^{2-}$), which rapidly converts the I_2 formed in Reaction 1 back to iodide ion:

$$I_2 + 2S_2O_3^{2-} \rightarrow 2I^- + S_4O_6^{2-} \quad \text{Reaction 2}$$

While Reaction 2 is said to be fast compared to Reaction 1, it can actually proceed no faster than the rate at which I_2 is produced, so its rate is controlled by the rate of Reaction 1. When the $S_2O_3^{2-}$ is completely consumed, then I_2 begins to accumulate. It combines with a starch indicator to form a dark blue complex, so the end point of the titration is determined by the time required for the solution to turn blue. The time it takes for the color to appear is called the period of the clock, or Δt, the time it takes for Reaction 2 to go to completion. The rate of $S_2O_3^{2-}$ disappearance is therefore given by its initial concentration divided by the clock period, or $[S_2O_3^{2-}]/\Delta t$. Because the rate of disappearance of I_2 in Reaction 2 is controlled by its rate of appearance in Reaction 1, the rate of both reactions can be determined from the value of $\frac{1}{2} \times [S_2O_3^{2-}]/\Delta t$. (Two $S_2O_3^{2-}$ ions are consumed for every I_2 consumed.) Actually, the I_2 in solution combines with iodide ion to form triiodide ion, I_3^-, which is pale yellow. Starch is added to form a blue color that is much easier to observe. The reaction between the triiodide ion and starch is not shown to keep the equations simple, but adding it would not change the conclusions.

Typically, reactions occur in more than one step. The slow step determines the rate. The "clock reaction" consists of a slow step followed by a fast step:

$$\text{Slow} \quad S_2O_8^{2-} + 2I^- \rightarrow 2SO_4^{2-} + I_2 \quad \text{Reaction 1}$$

$$\text{Fast} \quad I_2 + 2S_2O_3^{2-} \rightarrow 2I^- + S_4O_6^{2-} \quad \text{Reaction 2}$$

The overall reaction will be observed in lab by using the reaction of iodine and starch, which forms a blue complex. The appearance of the blue complex is an indication that the reaction is complete. As mentioned above, the thiosulfate ion ($S_2O_3^{2-}$) is the limiting reactant in the clock reaction. The I^- (or I_3^-) that is needed to react with the starch, indicating the end point of the reaction, does not begin to form until all of the $S_2O_3^{2-}$ is gone. The rate of the reaction can therefore be determined by measuring the amount of time required for the blue color to appear.

The Rate Law

The rate law for the reaction with respect to the concentrations of each reactant is expressed as:

$$rate = \frac{-d[I^-]}{dt} = \frac{-d[S_2O_8^{-2}]}{dt} = k[I^-]^m[S_2O_8^{-2}]^n$$

To determine the value of the coefficients m and n in the rate law, we must first determine the rate of the reaction with respect to each reactant. In other words, what effect does the concentration of each reactant have on the overall reaction rate? The computation of the rate is complicated by the fact that the rate is not constant but changes throughout the reaction. But there is a way we can simplify this computation. For the clock reaction, it is possible to choose concentrations of reactants such that the rate remains constant until one reactant, the limiting reagent, is entirely consumed. Under these conditions, the calculation of the rate becomes the following:

$$rate\ of\ reaction = \frac{initial\ amount\ of\ limiting\ reactant}{time\ of\ observation}$$

If the number of moles of limiting reagent put into the reaction solution at the start of the reaction is known, as well as the total volume of the reaction solution, the initial concentration can be calculated. The time (in seconds) required for the complete reaction (at constant rate), is then recorded in lab, and the rate of the reaction can be determined.

To determine the rate law, the rate order with respect to each reactant in the slow step will be determined. If the peroxydisulfate ion $(S_2O_8^{2-})$ is used in large excess so that its concentration remains essentially constant throughout the reaction and only the KI concentration is varied, we can determine how the rate of the reaction varies with respect to iodide ion. If we next hold the concentration of KI in excess and vary the peroxydisulfate ion concentration, we can determine its contribution to the rate as well.

Colorimetric Determination

When has the limiting reagent been consumed? Iodine is formed in the slow step but is removed rapidly by the reaction with the limiting reagent $S_2O_3^{2-}$. When all of the $S_2O_3^{2-}$ is used up, yellowish I_2 begins to accumulate. (The other reactants and products are colorless.) However, the iodine color is faint and difficult to observe. Sensitivity can be improved by using starch as an indicator. In the presence of a suitably prepared starch solution, I_2 forms a deep blue starch-tri-iodide complex. In effect, the starch indicator amplifies the color of the iodine. At exactly the time when all of the thiosulfate $(S_2O_3^{2-})$ has been consumed, the solution will change from colorless to blue. We are measuring the rate of slow step because the reaction shown in the fast step is too fast.

Determining the Rate Law

Once we have determined the rate of the reaction, we want to determine the rate law for the reaction. A rate law of a reaction is a mathematical expression relating the rate of a reaction to the concentration of either reactants or products. The rate law may be theoretically determined from the rate-determining step (slow step) of the reaction mechanism. Many chemical reactions actually require a number of steps to break bonds and form new ones. The rate law of a reaction must be proven experimentally by looking at either the appearance of products or the disappearance of reactants.

The following examples show how the rate law can be determined for a reaction similar to that of the iodine clock.

Example Problem

Consider the following overall chemical reaction:

$$2NO(g) + O_2(g) \Leftrightarrow 2NO_2(g)$$

Here is an example of a rate law for the above reaction:

$$k = [NO]^2[O_2]$$

In this rate law, the rate of disappearance of NO and O_2 is proportional to the concentrations of NO and O_2, where k is a rate constant (a proportionality constant). The exponent associated with each concentration term is referred to as the *order*. The sum of the individual orders gives the overall order of the reaction. In the above rate law, the order for NO is 2 and is often referred to as second order. The order for O_2 is 1 and is often referred to as first order. The overall order is 3 (third order). It is important to note that the coefficients found in the balanced equation are not necessarily related to the exponents found in the rate law.

Let's look at a set of experimental data from which the rate law may be deduced. In this case, we will examine the rate in terms of the disappearance of reactants. Experimentally, one compiles the reaction rates relative to initial concentrations of reactants:

	Rate M/s	[NO] M	[O$_2$] M
Experiment 1	1.20×10^{-8}	0.10	0.10
Experiment 2	2.40×10^{-8}	0.10	0.20
Experiment 3	1.08×10^{-7}	0.30	0.10

Where the rate is determined by:

$$rate = \frac{d[reactant]_i}{dt}$$

where [reactant]$_i$ is the initial concentration of the limiting reactant and dt is the time it takes for the reaction to run to completion.

The rate law can be determined using two methods. The first method is by examining the experimental data. Pick two trials to compare in which the concentration of one reactant is changed while the concentration of the other reactant is held constant. We will determine the order for the reactant whose concentration has changed. By keeping the concentration of the other reactant constant, we ensure that the rate will not be affected by that reactant.

Let's begin by choosing the results for Trial 1 and Trial 2. In these two trials, the concentration of NO is held constant while the concentration of O_2 has been doubled. Also, note that doubling the concentration of O_2 doubles the rate of the reaction. This may be mathematically expressed as

$$\frac{[O_2]_f^x}{[O_2]_i^x} = \frac{0.20}{0.10} = 2 \text{ and } \frac{\text{rate final}}{\text{rate initial}} = \frac{2.40 \times 10^{-8}}{1.20 \times 10^{-8}} = 2$$

or

$$[O_2]^x = \text{rate}$$

Mathematically, for the rate to be directly proportional to the change in concentration x (the reaction order) must equal 1. This makes the order for O_2 first order.

The same procedure is used to determine the order of NO. Pick two trials in which the concentration of O_2 is held constant while the concentration of NO is varied. These criteria are met with Trial 1 and Trial 3. The concentration of O_2 is held constant while the concentration of NO has tripled. Under these conditions, the rate of the reaction is increased nine fold. This can be expressed mathematically as:

$$\frac{[NO]_f^x}{[NO]_i^x} = \frac{0.30}{0.10} = 3 \text{ and } \frac{\text{rate final}}{\text{rate initial}} = \frac{1.08 \times 10^{-7}}{1.20 \times 10^{-8}} = 9$$

or

$$[NO]^x = \text{rate}$$

For the rate to have increased by a factor of nine when the concentration was tripled, x must be 2. This makes the order for NO second order. We are now able to write the rate law: rate = $k[NO]^2[O_2]$, where k is the rate constant.

To solve for the rate constant, choose any experimental trial and substitute the values for the rate and concentrations of NO and O_2. Trial 1 data will be arbitrarily chosen in this case:

$$k = \frac{\text{rate}}{[NO]^2[O_2]} = \frac{1.2 \times 10^{-8}\,M\!/\!s}{(0.10\,M)^2(0.10\,M)} = 1.2 \times 10^{-5}\,M^{-2}s^{-1}$$

Thus the rate law becomes: rate $= 1.2 \times 10^{-5}\,M^{-2}\,s^{-1}[NO]^2[O]$

The rate law for the above reaction can also be determined graphically by expressing the rate law in a straight-line form, which is what you will be doing in this experiment.

$$\textbf{rate} = \textbf{\textit{k}}[\textbf{NO}]^x[\textbf{O}_2]$$

By rearranging the rate law, taking the log of each side, and assuming one concentration is a constant:

$$k = \frac{\text{rate}}{[NO]^x[O_2]} \quad \text{Multiply both sides by } [O_2]$$

$$k[O_2] = \frac{\text{rate}}{[NO]^x} \quad \text{Take the log of both sides}$$

$$\log k[O_2] = \log \text{rate} - x \log [NO] \quad \text{Rearrange}$$

$$\log \text{rate} = x \log [NO] + \log k[O_2]$$

Graphing the log of the rate (y-axis) versus the log of the [NO] (x-axis) should result in a straight line graph with a slope equal to x, the reaction order for NO. This same process can be done for both reactants, simply by holding the concentration of one reactant constant each time.

Once the rate orders have been determined, the rate constant k can be determined by substituting the data for each run and solving for k. If the data is good, the k values should all be reasonably close, and the average k for all the runs can be reported.

Temperature Effects on Rate

In this experiment, we will also collect data at several temperatures, which will allow us to calculate the reaction's activation energy. The mathematical relationship between temperature and activation energy is given by this variation of the Arrhenius equation:

$$\ln k = \frac{-E_a}{R}\left(\frac{1}{T}\right) + \ln A$$

In this equation, A, a constant called the frequency factor, is related to the frequency of collisions and the probability that the collisions are favorably oriented; R is the universal gas constant, expressed in J/mol · K; E_a is the activation energy in joules, T is the temperature in Kelvin; and k is the rate constant. Since this version of the Arrhenius equation is in the form of a straight line equation, a plot of ln (k) on the y-axis versus $1/T$ on the x-axis will give a straight line. The slope of the line will equal $-E_a/R$, and the y-intercept equals ln A. Therefore, we can use the Arrhenius equation to determine the activation energy and frequency factor for the iodine clock reaction.

The chemical kinetics of the iodine clock can be studied by simply measuring the rate of the reaction with varying concentrations of reactants and at several temperatures. Once the rate of the reaction is determined, one can mathematically determine the rate constant for the reaction at the given conditions and the amount of energy it takes to activate the reaction.

Procedure

SAFETY NOTES: Use good chemical safety techniques as usual. Make sure all waste goes into the containers provided and not down the sink. Wear your goggles!

GENERAL INSTRUCTIONS: You will work with a partner for this lab. Before starting, determine the room temperature. Careful labeling of reagents and the graduated cylinders used to measure them is essential to the success of this experiment.

Part I: Kinetics at Room Temperature

Use a clean 250 mL Erlenmeyer Flask to combine the reagents for each run (1 through 5 and 8) listed in the table in the prelab. Use a different graduated cylinder to measure each reagent. The reagents should be added one at a time in the order they are listed, from left to right.
Be ready to start timing the reaction when you add the final reagent, $(NH_4)_2S_2O_8$.

Stop timing each reaction when the first blue color appears. Record the elapsed time in your notebook.

Initial Concentrations of Reactants					
Run #	KI 0.20 M	$Na_2S_2O_3$ 0.00175 M	Water	Starch	$(NH_4)_2S_2O_8$ 0.050 M
1	30 mL	20 mL	0 mL	3 drops	30 mL
2	20 mL	20 mL	10 mL	3 drops	30 mL
3	10 mL	20 mL	20 mL	3 drops	30 mL
4	30 mL	20 mL	10 mL	3 drops	20 mL
5	30 mL	20 mL	20 mL	3 drops	10 mL
6	30 mL	20 mL	0 mL	3 drops	30 mL
7	30 mL	20 mL	0 mL	3 drops	30 mL
8	30 mL	10 mL	10 mL	3 drops	30 mL

Part II: Kinetics at Elevated Temperature

For run 6, collect a hot plate from the front counter. Turn it on and set it to 10 to 15 °C above room temperature. Place a 400 mL beaker containing 200 mL of water on the hot plate.

Put all of the reagents except for the $(NH_4)_2S_2O_8$, in an Erlenmeyer flask and place it in the water bath to equilibrate.
Place the $(NH_4)_2S_2O_8$ in a large test tube and place it in the water bath as well.
Leave the reagents in the bath until they both stabilize to the temperature of the bath water. Then add them together and record the time as before.

Part III: Kinetics at Lowered Temperature

For run 7, collect a plastic tub and ice. Add water and salt to make an ice bath.

Place the reagents in the ice bath in the same way you placed them in the hot water bath.
Wait for them to equilibrate, combine them, and record the reaction time as before.

Report Contents and Questions

The purpose should be several sentences summarizing the objectives of the experiment and how those objectives will be assessed.

The procedure section should reference the lab manual and note any changes made to the experiment.

The data section should include the following completed table:

Run Number	[KI] (M)	[Na$_2$S$_2$O$_3$] (M)	[(NH$_4$)$_2$S$_2$O$_8$] (M)	Time (sec)	Rate of Reaction (M/sec)	Temp. (K)	k (determine units)
1							
2							
3							
4							
5							
6							
7							
8							

To determine the [KI], determine the number of moles in the volume of KI you are using in each trial and divide that amount by the total volume for each run. (Hint: 20 drops = 1 mL.) Repeat these calculations to determine the [Na$_2$S$_2$O$_3$] and [(NH$_4$)$_2$S$_2$O$_8$]. To calculate the rate of the reaction, divide the [Na$_2$S$_2$O$_3$] by the time.

Next, to determine k, fill in the table below and make two graphs.

Run Number	log[kI] (M)	log[(NH$_4$)$_2$S$_2$O$_8$] (M)	log Rate of Reaction (M/sec)
1			
2			
3			
4			
5			
6			
7			
8			

One graph will have the log of the rate on the y-axis and the log of [KI] on the x-axis only using runs 1 through 3 the slope of this graph will give the value of m in the following equation:

$$\log \text{rate} = m \log k[\text{I}^-] + n \log[\text{S}_2\text{O}_8{}^{2-}]$$

To find the value of n you will do the same thing but instead of graphing the log of [KI], graph the log of $[(NH_4)_2S_2O_8]$ using runs 1. Now that you have m and n, you can substitute the rate, [KI], and $[(NH_4)_2S_2O_8]$ into the rate law:

$$\text{rate} = k[I^-]^m[S_2O_8^{2-}]^n$$

and solve for k for each of the eight runs. The units for k are determined by the values of m and n. The values for m and n should be whole numbers so be sure to round your answers correctly.

Finally, determine the values of E_a and A, the activation energy and frequency factor for this reaction. These are determined by preparing a graph of ln k versus $1/T$ for trials 1, 6, and 7. The slope of this graph is equal to $-E_a/R$ and the y-intercept is ln A. Take note: the equations for determining E_a use the natural logarithm, ln.

In the calculation section include the three graphs explained above. All graphing guidelines apply. Also include sample calculations for the following: [KI], $[Na_2S_2O_3]$, $[(NH_4)_2S_2O_8]$, rate, k, E_a, and A.

The conclusion section should be several paragraphs containing all of the values obtained, including the rate law. This section should also include a comparison of the rate constants at the same temperature; a well developed discussion of the rate constants for all runs, including both similarities and differences; and an explanation of the value for E_a. As always, discuss any possible errors in the experiment.

Answer the following questions:

1) The *kinetic molecular theory* (KMT) was introduced earlier in the chapters of the textbook which accompanies this lab manual. Use KMT to explain how temperature affects the rates of chemical reactions. Be detailed in your explanation.

Experiment 18
Laboratory Preparation

Name: _____ Date: _____

Instructor: _____ Sec. #: _____

Show all work for full credit.

1) Read the background, procedure, and report sections of the lab experiment carefully and develop a hypothesis of what information you expect to gain from the completion of the lab experiment.

2) Create any and all tables you might need to collect data during the experiment and then transfer those tables into your lab manual for use during the lab.

Experiment 18
Prelaboratory Assignment

Name: _____ Date: _____

Instructor: _____ Sec. #: _____

Show all work for full credit.

1) For KI, $Na_2S_2O_3$, and $(NH_4)_2S_2O_8$, calculate the initial concentration (M) based on the table of solution concentrations and volumes given in the procedure section. *Copy the completed table into your lab notebook for later use*. The shaded columns will be completed in lab.

Run Number	[KI] (M)	[Na$_2$S$_2$O$_3$] (M)	[(NH$_4$)$_2$S$_2$O$_8$] (M)	Time (sec)	Rate of Reaction (M/sec)	Temp. (K)	K (determine units)
1							
2							
3							
4							
5							
6							
7							
8							

2) A first order chemical reaction has a rate constant of 9.1×10^1 s^{-1} at 12 °C and a rate constant of 4.1×10^2 s^{-1} at 36 °C.

Calculate the activation energy, E_a, for the reaction.

3) In the iodine clock reaction, you are measuring the rate of the reaction of iodide ion (I^-) with peroxydisulfate ion ($S_2O_8^{2-}$):

$$S_2O_8^{2-} + 2I^- \rightarrow 2SO_4^{2-} + I_2 \qquad \text{(Reaction 1)}$$

The rate of the reaction is measured by determining the amount of I_2 formed in a particular time, or $\Delta[I_2]/\Delta t$. The I_2 is measured by titrating it with a fixed amount of thiosulfate ion ($S_2O_3^{2-}$), which rapidly converts the I_2 formed in Reaction 1 back to iodide ion:

$$I_2 + 2S_2O_3^{2-} \rightarrow 2I^- + S_4O_6^{2-} \qquad \text{(Reaction 2)}$$

(While Reaction 2 is said to be "fast" with respect to Reaction 1, it actually can proceed no faster than the rate at which I_2 is produced, so its actual rate is controlled by the rate of Reaction 1.)

When the $S_2O_3^{2-}$ is completely used up, then I_2 begins to accumulate. It combines with starch to form a dark blue complex, so the end point of the titration occurs when the solution turns blue. The time it takes for the blue color to appear is called the period of the clock, or Δt, the time it takes for Reaction 2 to go to completion. The rate of $S_2O_3^{2-}$ disappearance is therefore given by its initial concentration divided by the clock period, or $[S_2O_3^{2-}]/\Delta t$. Because the rate of disappearance of I_2 in Reaction 2 is controlled by its rate of appearance in Reaction 1, the rate of both reactions can be determined from the value of $\frac{1}{2} \times [S_2O_3^{2-}]/\Delta t$. (Two $S_2O_3^{2-}$ ions are consumed for every I_2 consumed.)

(Actually, the I_2 in solution combines with iodide ions to form triiodide ion, I_3^-, which is the species that forms the blue color with starch. That step is not shown to keep the equations simple, but adding it would not change the conclusions.)

The graphs to the right illustrate what is happening to the concentration of each of the species involved in these two reactions. Show you understand the reaction by identifying the species associated with each of the lines in the graphs.

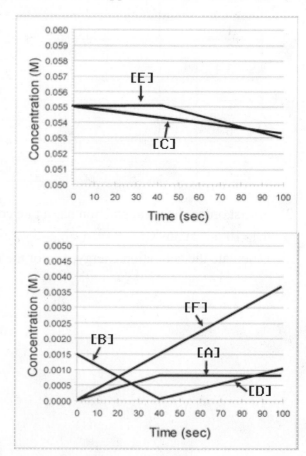

Initial conditions:

$[S_2O_8^{2-}] = [I^-] = 0.055$ M
$[S_2O_3^{2-}] = 0.0015$ M

$[SO_4^{2-}] =$ _____ $[I^-] =$ _____

$[I_2] =$ _____ $[S_2O_3^{2-}] =$ _____

$[S_4O_6^{2-}] =$ _____ $[S_2O_8^{2-}] =$ _____

Experiment 19

Photometric Determination of an Equilibrium Constant

Introduction

This experiment introduces the concept of equilibrium using the formation of a complex ion $Fe(SCN)_x^{(3-x)}$. By determining the exact concentration of each species in equilibrium, the equilibrium constant, K, can be derived. There are several reasons why a scientist would want to determine the equilibrium constant. K can provide information about solubility, acidity, and even structural conformations. Primarily, however, a chemist's goal is to determine how to best produce a desired material. When the K value is very large (greater than 1), the reaction is said to lie to the right (of the arrow), indicating that the amount of product is greater than the amount of reactants; values less than 1 indicate a higher concentration of reactants than products, and values close to 1 indicate similar amounts of products and reactants. If a scientist can develop reaction conditions (temperatures, pressures, concentrations) that generate a $K > 1$, then more of a desired product will be formed. Obviously, making more product with the same amount of reactant(s) is a goal of any manufacturing process.

To determine the value of K experimentally, a method by which the concentrations of all species in the reaction can be determined must be used. Luckily, in the reaction we will be investigating, the complex ion product, $Fe(SCN)_x^{(3-x)}$, is colored. This will afford us the use of a technique called absorption spectroscopy.

Absorption spectroscopy has already been used in the solutions and colloids lab. Unlike the use of spectroscopy in that lab, our purpose here is not only to create a calibration curve but also to determine the concentration of $Fe(SCN)_x^{(3-x)}$ by use of Beer's law, $A = \varepsilon bc$. Quantitative Measurements will then be used to derive the equilibrium concentrations of the other species in the reaction.

The use of a Mohr or graduated pipette is a technique introduced in this lab. It is very important that the volumes of the solutions used be accurate, and the Mohr pipette will allow us to minimize error when creating the needed solution concentrations.

Finally, you will again be making use of Excel (or a similar spreadsheet program) in this experiment. Not only is a graph needed in this lab, but you will also be making repetitive calculations that could be quite arduous if done by hand. Excel can simplify the calculations since minor changes in a single cell can be translated into every other cell (see Appendix B).

Background

In its most abstract form, equilibrium is defined as a state in which a particular system at a given temperature has allowed its energy to be distributed in the statistically most probable manner. In simpler terms, equilibrium is reached when the forces, influences, and reactions in a system balance each other out so that there is no net change. For example, your body is said to be in thermal equilibrium when there is no net heat exchange between it and your surroundings.

In this experiment, you will be introduced to chemical equilibrium. The generalized chemical equation for a reversible reaction where the capital letters (A, B, C, and D) represent the species involved in the reaction and the lower case letter (a, b, c, and d) correlate to the numerical coefficients indicating how many moles of each reactant and product are involved is

$$aA(aq) + bB(aq)... \Leftrightarrow cC(aq) + dD(aq)...$$

The Equilibrium Constant

Perhaps the most important component of equilibrium is the equilibrium constant. This value, expressed as K_c when concentrations are used and K_p when partial pressures are used, reveals the position of equilibrium. K_c is generally written as a ratio of the concentration of the products, with their appropriate powers, to the concentration of the reactants, with their respective powers. For our example, the equilibrium constant for the above reaction is:

$$K_c = \frac{[C]^c[D]^d...}{[A]^a[B]^b...}$$

In the previous paragraph, it was mentioned that the magnitude of the equilibrium constant reveals the position of equilibrium. Recalling that the constant is a ratio of products to reactants, it is rather intuitive that a small value of K indicates that the amount of product (C and D) is relatively small compared to the amount of reactant (A and B). In other words, when K is small, the reverse reaction is dominant. The equilibrium expression, can also indicate just how the equilibrium will shift when reactants or products are added to or removed from the system.

Example Problem

To further clarify the significance of the equilibrium constant and to familiarize you with some of the calculations it involves, look at the simple reaction shown below:

$$2NH_3(g) \Leftrightarrow N_2(g) + 3H_2(g)$$

Assume that at a particular temperature, 6.00 mol of NH_3 is introduced into a 3.00-L flask and allowed to reach equilibrium. At equilibrium, only 3.00 mol of NH_3 remained. Using this information, we can determine the equilibrium constant.

To begin this problem, an expression for the equilibrium constant must be written:

$$K = \frac{[N_2][H_2]^3}{[NH_3]^2}$$

Since the number of moles of NH_3 remaining at equilibrium and a volume are given, it is easiest to deal with concentrations as opposed to partial pressures. Take the 3.00 mol of NH_3 remaining at equilibrium and divide by the volume of the vessel to obtain the equilibrium concentration of ammonia.

$$[NH_3]_{eq} = \frac{3.00 \text{ mol}}{3.00 \text{ L}} = 1.00 \text{ M}$$

The equilibrium expression now looks like this:

$$K_c = \frac{[N_2][H_2]^3}{[1.00]^2}$$

However, there are still two unknowns, $[N_2]$ and $[H_2]$, whose concentrations must be obtained to calculate the equilibrium constant. You can always use an ICE table to calculate the equilibrium concentrations based on the equilibrium concentration of NH_3. However, looking back at the coefficients for each species in the reaction, it is apparent that the concentration of NH_3 is twice that of N_2 and two-thirds that of H_2. Mathematically, the respective concentrations can also be calculated as shown below:

$$[N_2]_{eq} = \tfrac{1}{2}[NH_3] = \tfrac{1}{2}[1.00 \text{ M}] = 0.500 \text{ M}$$
$$[H_2]_{eq} = \tfrac{3}{2}[NH_3] = \tfrac{3}{2}[1.00 \text{ M}] = 1.50 \text{ M}$$

Using the concentrations of all the species at equilibrium, the value of the equilibrium constant becomes:

$$K = \frac{[0.50][1.50]^3}{[1.00]^2} = 1.69$$

With an equilibrium constant of ~1.7, the reaction is slightly favored in the forward reaction. In other words, for this particular reaction at the given temperature, the concentration of the products is slightly more than that of the reactants.

The Experiment

In this experiment, the equilibrium concentration of a complex ion is to be determined by use of absorption spectroscopy. The first part of the experiment, however, requires the determination of the extinction coefficient, ε. As in the solutions and colloids lab, the Beers–Lambert relationship between concentration and absorption is used to create a calibration graph. As long as one concentration in the reaction remains constant, the graph will be (hopefully) linear with a slope equal to ε.

To generate a calibration curve, the absorbance of a series of solutions with different concentrations of the colored species is measured at a specific wavelength. A plot of absorbance (*y*-axis) versus [KSCN] (*x*-axis) is created to generate a linear graph corresponding to the equation $A = \varepsilon bc$, where the extinction coefficient is equal to the slope. This extinction coefficient is then used in a later series of reactions to calculate actual concentrations of the colored complex.

Determining the Equilibrium Constant, *K*

As with the example problem above, the first step in determining the equilibrium constant is writing a balanced chemical equation.

$$Fe^{3+}(aq) + xSCN^-(aq) \Leftrightarrow Fe(SCN)_x^{(3-x)}(aq)$$

The reaction above poses some interesting challenges because the exact chemical formula of the product is unknown. The subscripts and coefficients represented with an *x* shown in the reaction above could be 1, 2, or 3, yielding three different balanced reactions, respectively:

$$Fe^{3+}(aq) + SCN^-(aq) \Leftrightarrow Fe(SCN)^{2+}(aq) \qquad \text{when } x = 1$$

or

$$Fe^{3+}(aq) + 2SCN^-(aq) \Leftrightarrow Fe(SCN)_2^+(aq) \qquad \text{when } x = 2$$

or

$$Fe^{3+}(aq) + 3SCN^-(aq) \Leftrightarrow Fe(SCN)_3(aq) \qquad \text{when } x = 3$$

To determine the equilibrium constant, the equilibrium concentrations of all the products and reactants must be determined several times, starting with different initial concentrations of reactants. Although this seems complicated, all it really means is a little more work. Once the equilibrium concentration of $Fe(SCN)_x^{(3-x)}(aq)$ is experimentally determined based on the reactant combinations above, calculations of the equilibrium concentrations of the reactants are completed in the exact same manner as shown in the example problem. The calculations simply must be completed three times—once for each possible *x* value. The use of a spreadsheet greatly simplifies these calculations.

The only question that will remain at that point is which *x* value is the correct one. The answer to that will be readily apparent when you calculate the average and standard deviations of the equilibrium constant for each run. The value that produces the most consistent value of *K* (smallest standard deviation) will be the correct value to assume for the reaction.

Procedure

SAFETY NOTES: The solutions being used are acidified. Use all appropriate precautions. Wear goggles at all times and wash your hands immediately if your skin begins to itch or tingle (i.e., if you spill acid on yourself).

Part I: Making Solutions to Determine the Extinction Coefficient

In two clean, dry 100 mL beakers, collect 50 mL of 0.200 M acidified $Fe(NO_3)_3$ and 0.001 M KSCN.

Using a Mohr pipette, create the solutions A through E described in the prelab. Be very careful not to contaminate your stock solutions. If a rust color forms in the pipette, it is contaminated. Clean it and remake the solution you were working on.

Note: Each solution should have the exact same total volume.

Part II: Measuring the Absorbance

Collect a cuvette. Make sure it is clean and dry.

Create a blank of 0.200 M $Fe(NO_3)_3$. Set the spectrometer to a wavelength of 480 nm.

Rinse the cuvette with the blank and then fill the cuvette with blank and zero the spectrometer.

Rinse the cuvette with the first solution and then refill it with the solution to be tested. Measure the absorbance and record it in your notebook.

Repeat the process until all the samples are measured.

Part III: Making Solutions to Determine the Equilibrium Constant

In clean, dry 100 mL beakers, collect 50 mL of 0.002 M acidified $Fe(NO_3)_3$ and 0.002 M KSCN.

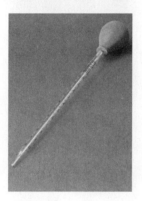

Using a Mohr pipette, create the solutions F through J described in the prelab. Be very careful not to contaminate your stock solutions. If a rust color forms in the pipette, it is contaminated. Clean it and remake the solution you were working on.

Note: Each solution should have the exact same total volume.

Part IV: Measuring the Absorbance

Collect a cuvette. Make sure it is clean and dry.

Create a blank of 0.002 M $Fe(NO_3)_3$. Set the spectrometer to a wavelength of 480 nm.

Rinse the cuvette with the blank and then fill the cuvette with blank and zero the spectrometer.

Rinse the cuvette with the first solution and then fill it with the solution. Measure the absorbance and record it in your notebook.

Repeat the process until all the samples are measured.

Solution	mL 0.200 M Fe(NO$_3$)$_3$	mL 0.002 M KSCN	mL 0.1 M HNO$_3$	Absorbance
A	5	1	14	
B	5	2	13	
C	5	3	12	
D	5	4	11	
E	5	5	10	

Solution	mL 0.002 M Fe(NO$_3$)$_3$	mL 0.002 M KSCN		Absorbance
F	3	7		
G	4	6		
H	5	5		
I	6	4		
J	7	3		

Report Contents and Questions

The purpose should be several well-constructed sentences dealing with the major topics of this experiment. Include both conceptual and technique-based topics and your criteria for determining success. The procedure section should cite the lab manual and also note any changes made to the procedure in the lab manual.

The data section for this experiment will include two sections: one for the calibration data (determination of the extinction coefficient) and one for the equilibrium data (determination of the equilibrium constant). For the calibration data, include a table with the solutions A through E, the volume of Fe^{3+}, the volume of KSCN, the total volume, the $[Fe^{3+}]$, the [KSCN], and the absorbance values for each experimental run.

For the equilibrium data (solutions F through J), consider the following chemical reaction and equilibrium expression:

$$Fe^{3+}(aq) + xSCN^-(aq) \Leftrightarrow Fe(SCN)_x^{(3-x)}(aq)$$

$$K_c = \frac{[Fe(SCN)_x^{(3-x)}]}{[Fe^{3+}][SCN^-]^x}$$

where x is 1, 2, or 3.

To determine the value of x, complete the following table for each x value and then determine the set of data with the most constant K value. The data with the most constant K value is the one with the correct value for x.

Complete the following table for $x = 1$, 2, and 3:

Soln	Initial Volume of Fe^{3+} (mL)	Initial Volume of KSCN (mL)	Total Volume (mL)	Initial $[Fe^{3+}]$ (M)	Initial $[SCN^-]$ (M)	abs	$[Fe(SCN)_x^{(3-x)}]$ (M)	Eq $[Fe^{3+}]$ (M)	Eq $[SCN^-]$ (M)	K
F										
G										
H										
I										
J										

The initial concentrations are found the same way as above. The absorbance value was recorded in lab. The $[Fe(SCN)_x^{(3-x)}]$ can be found using Beer's law:

$$A = \varepsilon b c$$

where A is the absorbance value, ε is the extinction coefficient from the slope of your calibration graph in Part I, b is the path length of the spectrometer (1 cm), and c is the concentration of $[Fe(SCN)_x^{(3-x)}]$.

To determine the equilibrium concentrations of $[Fe^{3+}]$ and $[SCN^-]$, use the following two equations, where $x = 1, 2,$ or 3:

$$[Fe^{3+}(aq)]_{eq} = [Fe^{3+}(aq)]_{init} - [Fe(SCN)_x^{(3-x)}]_{eq}$$

and

$$[SCN^- Fe^{3+}(aq)]_{eq} = x[SCN^-(aq)]_{init} - [Fe(SCN)_x^{(3-x)}]_{eq}$$

After determining the concentrations at equilibrium, determine K_c for each x value and reaction solution using the following equation:

$$K_c = \frac{[Fe(SCN)_x^{(3-x)}]}{[Fe^{3+}][SCN^-]^x}$$

Finally, for each x value or set of data, calculate the average K and the standard deviation in K. The x value with the smallest standard deviation in K is the correct value.

The calculation section, should include a calibration graph of the [KSCN] versus absorbance from Part 1. All graphing guidelines apply. The slope of this graph is the extinction coefficient used in Part II to determine the equilibrium concentration of $[Fe(SCN)_x^{(3-x)}]_{eq}$. Also be sure to include sample calculations for the concentrations used in the calibration graph and each calculation done for the equilibrium data.

In the conclusion section, include several paragraphs discussing the values determined for the extinction coefficient, the equilibrium constant K, and the value for x. Be sure to discuss the data that support these conclusions. Compare and contrast the values calculated when each of the three values for x are used. Do not forget to discuss any errors.

Answer the following questions:

1) How constant were your K values at room temperature? Explain any variation.

2) See if you can find literature and/or Internet references for the equilibrium constant for this equilibrium. Please cite the reference(s). How does your result compare with the one(s) you cited? Perhaps you will even find a similar experiment, with sample results, posted on the Web.

Experiment 19
Laboratory Preparation

_____

Name: _____ **Date:** _____

Instructor: _____ **Sec. #:** _____

_____

Show all work for full credit.

1) Read the background, procedure, and report sections of the lab experiment carefully and develop a hypothesis of what information you expect to gain from the completion of the lab experiment.

2) Create any and all tables you might need to collect data during the experiment and then transfer those tables into your lab manual for use during the lab.

Experiment 19
Prelaboratory Assignment

Name: _____ Date: _____

Instructor: _____ Sec. #: _____

Show all work for full credit.

1) You would like to determine the equilibrium constant for the binding of a ligand to a metal ion: $M^{3+} + L^- \Longleftrightarrow ML^{2+}$, where $K = [([ML^{2+}])/([M^{3+}][L^-])]$. The ML^{2+} complex has a unique absorbance in the visible region, so its concentration at equilibrium can be determined by measuring the absorbance of the solution at an appropriate wavelength. An ICE table can be used to determine the concentrations of M^{3+} and L^- from their initial concentrations and the concentration change in going to equilibrium. The following experiment is carried out first to determine the extinction coefficient of the complex at the wavelength chosen.

Calculate the complex concentration in each solution and then its extinction coefficient.

Solution	0.200 M Metal Ion (mL)	0.1 M HNO$_3$ (mL)	0.0011 M Ligand Ion (mL)	Complex Concentration (M)	Absorbance	Extinction Coefficient (M^{-1}cm^{-1})
A	5.0	4.0	1.0		0.113	
B	5.0	3.0	2.0		0.227	
C	5.0	2.0	3.0		0.337	
D	5.0	1.0	4.0		0.431	
E	5.0	0.0	5.0		0.562	

Plot the absorbance versus the solution concentration and determine the slope and intercept of the resulting line.

2) Now, you will determine the equilibrium constant, K, at several starting concentrations of M^{3+} and L^-. The following table summarizes your measurements.

Solution	0.0023 M Metal Ion (mL)	$[M^{3+}]_{init}$	0.0011 M Ligand Ion (mL)	$[L^-]_{init}$	Absorbance
F	3.0		7.0		0.286
G	4.0		6.0		0.312
H	5.0		5.0		0.308
I	6.0		4.0		0.276
J	7.0		3.0		0.221

Using an ICE table, calculate the equilibrium concentrations of ML^{2+}, M^{3+}, L^-, and K for each of the five solutions.

Solution	$[ML^{2+}] = x$	$[M^{3+}]_{eq}$	$[L^-]_{eq}$	$[ML^{2+}] = x$	K
F					
G					
H					
I					
J					

What value of K do you get for solution H?

What average value of K do you get?

Do you consider your values of K to be "constant"? Support your answer.

Experiment 20
Titration of 7UP®

Introduction

Two important concepts in chemistry are the focus of this experiment: titration and acid-base reactions. Titration is the method of determining the concentration of a solution by allowing a carefully measured volume to react with a standard solution of another substance, whose concentration is known. By adding a carefully measured volume of a base of known concentration to a carefully measured volume of acid of unknown concentration, or vice versa, and using your knowledge of stoichiometry and acid-base neutralization, you can determine the concentration of the unknown. In today's experiment, you will use this process to standardize a known volume of NaOH and determine the concentrations of unknown solutions of phosphoric acid and the citric acid content of 7UP.

The key phrase above for successful completion of this experiment is "your knowledge of stoichiometry and acid-base neutralization." Understanding these concepts is necessary for you to determine the unknown concentrations of three chemicals: sodium hydroxide, phosphoric acid and citric acid in 7UP. The other concept mentioned above, titration, is a new technique that you will use to make these determinations. Titration is a much-used process in chemistry and is an invaluable skill to master. It is one of the simplest and least expensive methods by which you can determine the concentration of an unknown solution, only requiring a titrant, a buret or other volumetric delivery device and a color indicator or pH meter.

The specific purpose of this experiment is to titrate a solution of a weak polyprotic acid (citric acid, $C_6H_8O_7$) and a stronger polyprotic acid (phosphoric acid, H_3PO_4) to determine their concentrations as well as their K_a values. The base (NaOH) used in these titrations will first be standardized using a commercially available standard HCl solution. The titration curves will be plotted and the concentrations and K_a values of each acid will be determined. By doing these titrations, you will have the opportunity to learn more about the chemical differences between weak and strong acids and also the meaning of polyprotic and how polyprotic compounds affects the outcome of a titration.

Background

The first in a series of similar experiments using titration to determine concentration, this investigation focuses on acid-base chemistry. In general terms, titrations utilize a known property of one solution to determine a similar property of an unknown solution. Specifically, these include acid-base titrations, potentiometric titrations (redox), complexometric (formation of a colored complex) titrations, and even titrations to determine specific concentrations of bacteria or viruses. For example, the alkalinity and acidity of water in streams and rivers is an important topic to environmental chemists. To observe such characteristics, they use the same technique you will learn in this experiment—acid-base titration.

Titration Set-up

The general setup of a titration is shown below (**Figure 20.1**). The base is placed in the buret so that a precise amount of solution can be added to the acid. The buret's precision is attributed to its graduations up and down the tube, making it one of the more expensive pieces of glassware you will use in the lab. The precision of the buret depends on reading it correctly: volumes delivered by a buret are read to the hundredth of a milliliter (0.01 mL). The knob on the buret is called a stopcock, and its sole purpose is to deliver the titrant to the solution below in a controlled manner.

Figure 20.1 Titration setup

Acid-Base Titrations

When an acid solution is titrated with a strong base such as NaOH, the initial pH of the solution is low. As base is added to the acidic solution, the pH gradually rises until the volume added is near the equivalence point, the point during the titration when equal molar amounts of acid and base have been mixed. Immediately before the equivalence point, the pH increases very rapidly and then levels off again immediately after the equivalence point with the addition of excess base (Figure 20.2). The equivalence point is located in the center of the vertical portion of this line. In this experiment, a carefully measured volume of HCl of a known concentration is titrated with NaOH of an unknown concentration. Since the buret allows us to determine the precise amount of base needed for neutralization, the precise concentration of the base can be calculated using $M_A V_A = M_B V_B$.

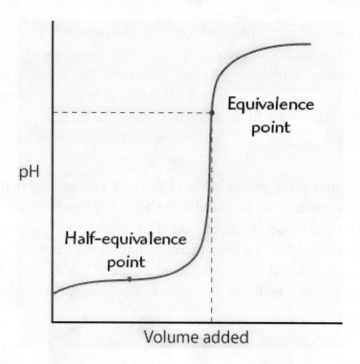

Figure 20.2 Titration curve of strong acid by strong base

Visualizing the end of a particular titration, specifically referred to as the end point, is essential to a successful titration. Indicators, often added in minute amounts to the solution of interest, are chemical compounds that undergo dramatic changes of color when a particular property of a solution is changed. Indicators are specific to the reaction being analyzed. The end point is the point in the titration where the indicator changes color, and the equivalence point is the point in the titration when the stoichiometric amount of titrant (equivalence volume) has been added such that the stoichiometric reaction has been completed. An indicator is generally chosen so that the end point is as close as possible to the equivalence point.

pH and pH Meters

The hydrogen ion concentration $[H_3O^+]$, expressed in terms of pH, is one of the most important properties of aqueous solutions, as it can control the solubility of various species, the formation of complexes, and even the kinetics of an individual reaction. To obtain precise data about the particular hydronium concentrations of the solutions in this experiment, and to clearly observe the change in pH at the equivalence point, a pH meter is used. In general, a pH meter measures the differences in electromotive force between two electrodes. A pH meter contains an electrode sensitive to the concentration of the hydrogen ion as well as one used solely as a reference.

For accurate measurements, it is necessary to calibrate the instrument using a buffer solution of approximately the same pH as the sample to be used. This calibration takes care of temperature effects and minor variations in the potential due to changes in the membrane. Your instructor will provide details regarding the calibration of the pH meters used in your laboratory.

The Experiment

There are a total of four acid-base titrations to be completed in this experiment: (1) standardize the titrant, a solution of sodium hydroxide (NaOH), using a known volume and concentration of hydrochloric acid (HCl); (2) repeat the standardization to confirm the concentration; (3) with the standardized solution of NaOH, determine the concentration of the relatively strong polyprotic acid, phosphoric acid (H_3PO_4); and (4) determine the concentration of the much weaker polyprotic acid, citric acid ($H_3C_6H_5O_7$), found in a sample of the soft drink 7UP.

Weak Acid Equilibria

A weak acid (HA) is one that does not fully dissociate in water. In other words, if the weak acid represented is allowed to ionize, as shown in the equation below, then a significant amount of HA will remain unionized.

$$HA(aq) + H_2O(l) \Leftrightarrow H_3O^+(aq) + A^-(aq)$$

At equilibrium, the dissociation of a weak acid is generally described by its acid-dissociation constant (K_a) and is mathematically represented as follows:

$$K_a = \frac{[H_3O^+][A^-]}{[HA]} \qquad (1)$$

In this investigation, you are asked to experimentally determine the acid-dissociation constants of both phosphoric acid and citric acid. With the knowledge that at equilibrium the concentration of the free hydronium ions (H_3O^+) is equal to the concentration of the conjugate base (A^-), if the concentration of either of these chemicals is determined experimentally, then stoichiometry can be used to determine the concentrations of the other components in the solution. Mathematically, the relationship for the reaction above is expressed as:

$$[HA]_{eq} = [HA]_{init} - [H_3O^+]_{eq} = [HA]_{init} - [A^-]_{eq}$$

From this logic, combined with the fact that pH is equal to the negative log of the hydrogen ion concentration, we can arrive at an expression for K_A by incorporating only the initial concentration of the weak acid, and the experimentally determined pH at the equivalence point.

$$K_a = \frac{[10^{-pH}]^2}{[HA]_{init} - [10^{-pH}]}$$

The acid-dissociation constant of a weak acid can also be obtained by another method. This method involves the half equivalence point, where just enough NaOH has been added to the weak acid to convert half of the acid to its salt. At this point, the concentration of the weak acid, [HA], is equal to the concentration of its conjugate base, [A^-]. Utilizing this fact, our generalized equilibrium expression equation (1) can now be defined as shown below because [A⁻] and [HA] can be canceled out of the expression.

$$K_a = [H_3O^+]_{eq} \qquad pK_a = pH \text{ at half-equivalence}$$

Overall, by performing all three of these titrations and plotting the pH versus volume of NaOH added, you can see how the pH of the solution changes as an acid, or a base, is added. In fact, if you are precise enough, you will get an idea of just how the shape of a titration curve can be influenced by both the concentration and nature of the acid or base. Further, by titrating both strong and weak acids, you will see different titration curves and how they change with the strength of the acid.

Polyprotic Acids

Both phosphoric acid and citric acid are triprotic acids, one form of polyprotic acid. This means that each of these acids has three ionizable hydrogens and thus three separate pK_a values, one for each dissociation. What is important to understand about a polyprotic acid dissociation is that at any point along the curve, there is some percentage of each acid form present in the solution. This means that unlike a monoprotic dissociation that is "all or nothing," the pH of a polyprotic acid solution is dependent on several forms of the acid. This also means that you should observe more than one inflection point in the titration curves.

Procedure

SAFETY NOTES: You will be handling strong acids and bases in this lab. Wash your hands immediately if you get some acid or base on your skin. Notify your instructor immediately regarding any spills that occur. You must wear goggles at all times.

Part 1: Standardization of NaOH

Obtain ~65 to 75 mL of NaOH and 30 to 35 mL of 0.01 M standard HCl. Clean a 50 mL buret and rinse it with the NaOH solution as follows. Add a few milliliters of the titrant and tip the buret to allow the solution to run almost all the way to the open end. Then rotate the buret slowly so that all surfaces contact the solution. Drain this solution into a waste beaker by opening the stopcock. Repeat this procedure two more times, using no more than 15 to 20 mL total. Fill the buret with the NaOH and open and close the stopcock to force air bubbles out of the tip. Use the buret clamp to place your buret on the ring stand.

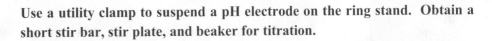

Use a utility clamp to suspend a pH electrode on the ring stand. Obtain a short stir bar, stir plate, and beaker for titration.

Calibrate the pH meter for data collection based on your instructor's directions.

Pipette 25.00 mL of standard 0.01 M HCl into a 250 mL beaker. Add a few drops of phenolphthalein to the beaker.

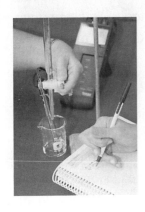

You are now ready to begin the titration. Read your buret to two decimal places, being sure that your eye is level with the meniscus. Always record the buret reading at the beginning and end of a titration. Your goal in this first titration will be to identify the pH region that surrounds the equivalence point(s). You will do this by consistently adding 2 mL of NaOH, constant stirring, and noting the change in pH after each addition. You will continue adding NaOH until you reach a point where the pH no longer changes (> pH 11).

NOTE: Stir the solution carefully. Avoid striking the pH probe. The probe is delicate and *expensive*.

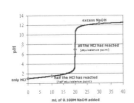

From the data collected above, you should be able to identify the regions of the titration where the pH changes very slowly and also the region of the equivalence point. You will now repeat the titration. The goal of this second titration is to generate enough data points to create an accurate and smooth curve.

Refill the buret if necessary, note the volume, and begin the second titration as previously instructed.

Repeat the steps above of the procedure using a 10.00 mL sample of phosphoric acid instead of the HCl. Add increments of ~0.2 mL until you reach a point where the pH no longer changes.

Repeat the steps of the procedure above using a 10.00 mL sample of 7UP instead of the HCl. Add increments of ~0.2 mL until you reach a point where the pH no longer changes.

Report Contents and Questions

The purpose should be several well-constructed sentences dealing with the major topics of this experiment. Include both conceptual and technique-based topics and your criteria for determining success. The procedure section should cite the lab manual and also note any changes made to the procedure in the lab manual.

The data section should include a balanced reaction for each of the acid dissociations being studied; there should be three sets of equations. These will include one equation with the reaction of NaOH and HCl, the three neutralization reactions for phosphoric acid by NaOH, and the three neutralization reactions for citric acid by NaOH. A table containing the concentration of NaOH determined from each of the two trials with standard HCl should be reported, along with the average value. The concentration and K_a values of phosphoric acid and citric acid should be reported. Finally, report the number of grams of citric acid found in a can of 7UP.

The calculation section should include four titration curves: two for HCl, one for phosphoric acid, and one for citric acid. Each graph (see Appendix B for a review of Excel) should be pH versus the volume of titrant added. All graphing guidelines apply. DO NOT add a trend line to these graphs; if you would like to add a line, a moving average would be acceptable. Sample calculations for determining the molarity of NaOH, phosphoric acid, and citric acid should be included. Sample calculations for K_a values: Recall that pH = pK_a at half equivalence point and that pK_a= –log (K_a). Theoretically, there should be one equivalence point for the HCl titrations and more than one for the titration of phosphoric acid and citric acid. Using the K_a values you calculate for each acid, calculate a percent error based on literature K_a values and cite your resource for these values. Finally, show a sample calculation of the number of grams of citric acid in a can of 7UP. NOTE: There are 240 mL of 7UP in a can. Make sure you present a sample calculation for all data presented in the data tables.

The conclusion should include several paragraphs with the following: the concentrations of HCl, NaOH, phosphoric acid, and citric acid; a discussion of the K_a values for the weak acids and the errors associated with them; a discussion of the differences in the titration curves for a weak acid and a strong acid; and a discussion of all possible errors.

Answer the following questions:

1) Are you able to determine all of the K_a values for citric acid? Why or why not?

2) Could this procedure be used for a cola drink instead of a non-cola drink? Why or why not?

Experiment 20
Laboratory Preparation

Name: _____ Date: _____

Instructor: _____ Sec. #: _____

Show all work for full credit.

1) Read the background, procedure, and report sections of the lab experiment carefully and develop a hypothesis of what information you expect to gain from the completion of the lab experiment.

2) Create any and all tables you might need to collect data during the experiment and then transfer those tables into your lab manual for use during the lab.

Experiment 20
Prelaboratory Assignment

Name: _____ Date: _____

Instructor: _____ Sec. #: _____

Show all work for full credit.
Copy your answers to problems 1 through 3 into your lab notebook for use when writing your lab report.

1) To standardize your NaOH solution, you titrate 20.0 mL of 0.0769 M HCl solution, measuring the pH after each small addition of NaOH. Below is the graph you obtain experimentally. The equivalence point occurs where there is a steep rise in pH.

What molar quantity of HCl is being titrated?

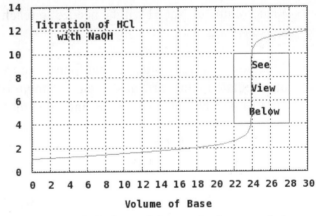

What molar quantity of NaOH has been added at the equivalence point?

What volume of NaOH solution has been added at the equivalence point?

Expanded view of the equivalence point:

What is the concentration of the NaOH solution?

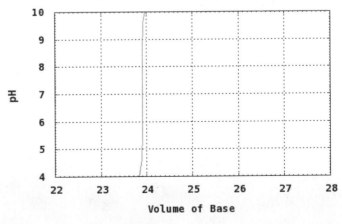

2) A triprotic acid requires three moles of base to neutralize it, and the protons are removed one at a time as follows:

$$H_3A + OH^- \leftrightarrow H_2A^- + H_2O$$
$$H_2A^- + OH^- \leftrightarrow HA^{2-} + H_2O$$
$$HA^{2-} + OH^- \leftrightarrow A^{3-} + H_2O$$

A titration curve, which plots the pH of the acid solution versus quantity of base that is added, can reveal several things about the nature of the acid. First, some or all of the equivalence points can be observed in the graph as places where there is a maximum in the slope of the curve.

As an example, you have a 40.0 mL solution of a triprotic acid, H_3A, with a concentration of 0.0588 M. You titrate it with a 0.250 M solution of NaOH. How many millimoles of H_3A are you titrating? (Give units.)

How many millimoles of NaOH are required to complete the titration? (Give units.)

What volume of NaOH will be needed to completely titrate the acid?

What volume of NaOH will be needed to reach the first equivalence point?

What volume of NaOH will be needed to reach the second equivalence point?

3) Citric acid is a weak organic acid found in citrus fruits and some soft drinks. It is an important intermediate in the citric acid cycle and therefore occurs in the metabolism of almost all living things. It is a triprotic acid, requiring three moles of OH^- to remove all the acidic protons. Titration of citric acid, as with all triprotic acids, occurs in a stepwise fashion in which one proton is removed at a time. The stepwise dissociation can therefore be represented by three equations, each with an acid-dissociation equilibrium constant.

$$H_3C_6H_5O_7 \leftrightarrow H_2C_6H_5O_6^- + H^+ \qquad (K_{a1})$$

$$H_2C_6H_5O_7^- \leftrightarrow HC_6H_5O_6^{2-} + H^+ \qquad (K_{a2})$$

$$HC_6H_5O_7^{2-} \leftrightarrow C_6H_5O_6^{3-} + H^+ \qquad (K_{a3})$$

The following graph shows how the concentration of each species changes when NaOH is added to a solution containing 100 mmoles of citric acid.

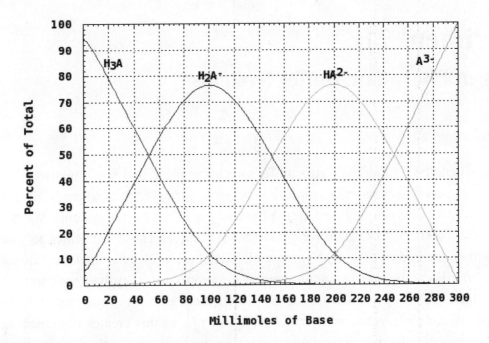

What is the molecular weight of citric acid?

What is the mass of citric acid used in this titration?

After addition of how much base does $[H_2C_6H_5O_7^-] = [C_6H_5O_7^{3-}]$?
After addition of how much base does $[H_2C_6H_5O_7^-] = [HC_6H_5O_7^{2-}]$?
After addition of how much base does $[H_3C_6H_5O_7] = [HC_6H_5O_7^{2-}]$?

Experiment 21

Hydrogen Phosphate Buffer Systems

Introduction

By now you should be starting to discuss buffer systems in class, and you might wonder if there is a purpose to the subject beyond what appears to be some very complicated math. Buffers are vital to the health of every living thing on Earth. Your body is full of buffer systems. We have learned previously that the H^+ concentration (or pH) of a solution can cause molecules to gain and lose protons based on their pK_a values. Proteins are biological molecules that make up the muscles and enzymes in your body. If proteins are not kept at the proper pH, they will not function properly. Enzymes, which are the protein catalysts by which most of your physiological processes are regulated, are rendered inoperative by the changing of pH, as this creates nonfunctional enzyme shapes. pH is probably the most important single thing the body regulates, since it affects so many other aspects of biochemistry. Proteins themselves often act as buffers; sodium bicarbonate and phosphate are others.

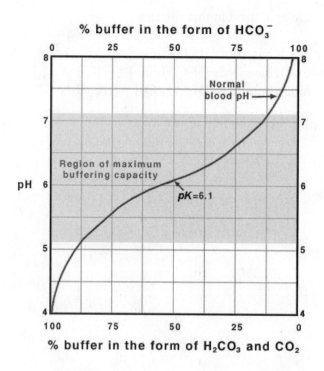

Your blood is buffered by the bicarbonate system. If disease overcomes the system, conditions known as acidosis or alkalosis (too much acid and too much base, respectively) occur. Mild symptoms of both are nausea, migraine, cramps, and fainting; severe forms can lead to coma and even death.

To study various cellular proteins in vitro, scientists must maintain the pH conditions that are found in the human body. This means that as a chemist or biochemist, you should be able to construct a buffer solution at a particular pH as needed.

This experiment is designed to give you practical experience in the calculations needed to create a buffer in the lab and to allow you to observe the buffering capacity of a phosphate buffer system. The repetitive nature of the experiment is designed to give you lots of practice with that "complicated" math.

Background

The pH of an acidic or basic solution is dependent on the concentration of hydrogen ions, H^+, also called free protons or **hydronium (H_3O^+)** ions:

$$pH = -\log[H_3O^+]$$

This concentration can be altered by adding substances that either react with or supply hydronium ions. Some solutions, however, exhibit the ability to resist pH changes—to a point—as additional acid or base is added. Such mixtures are said to be buffered, and the solutions are known as buffer solutions.

A buffer solution consists of an equilibrium system containing a weak acid (or base) and its conjugate. A simple example would be the mixture of acetic acid and sodium acetate. The acetate ion (CH_3COO^-) is the conjugate base of acetic acid. When only acetic acid is present, the amount of acetate ion in the equilibrium mixture is very small and can be determined from the K_a value and the molarity of the acid. The amount of acetate ion is sometimes described by the term *% dissociation*.

$$CH_3COOH + H_2O \Leftrightarrow CH_3COO^- + H_3O^+$$

$$K_a = 1.8 \times 10^{-5}$$

The small value of K_a tells us that this reaction has very little tendency to progress in the forward direction, that is, acetic acid is a weak acid and does not dissociate completely in water. Only about 1% of the acetic acid molecules actually donate protons to water at any given time. Thus, the equilibrium concentration of acetate ion is very small.

If additional acetate ion is added to acetic acid (in the form of sodium acetate or some other soluble acetate salt), H_3O^+ that have been formed by the small amount of dissociation will have a strong tendency to donate protons to the added acetate ions. The result of this "common ion effect" is that the H_3O^+ concentration decreases, and the pH of the mixture rises a little. We would predict exactly this behavior by LeChâtelier's principle. Because the concentration of free H_3O^+ is low to begin with, adding acetate will most likely leave an excess of acetate ion beyond that consumed by the reverse reaction. It will also most likely exceed the concentration that would be present due to the normal dissociation of acetic acid. This sets up the conditions for buffering: available supplies of the weak acid (or base) and the conjugate. When these supplies are of equal concentration, the buffer is said to be centered. This means it is able to absorb equal moles of acid or base before a significant change in pH occurs.

The key to understanding buffer behavior is to keep in mind that the pH of the mixture is related to the amount of free H_3O^+. If excess acid (e.g., some HCl) is added to the acetic acid–acetate buffer mixture, it readily donates protons to the acetate ion. As long as there are acetate ions available, added acid will be "absorbed" by this process and will not affect the concentration of H_3O^+ (or not much). Thus, the pH remains close to its original value. Note that this depends on having a large enough concentration of buffer so that some acetate will still remain after the addition of the acid.

Describing the behavior of the mixture when base is added is a bit more complicated. Hydroxide ion is a stronger base than acetate, so if sodium hydroxide is added to the buffer, H_3O^+ will donate protons to the hydroxide to form water. This loss of H_3O^+ upsets the equilibrium, and the acetic acid will donate protons to water in order to replace some of the lost H_3O^+. This nearly restores the original pH as long as there is a sufficient concentration of acetic acid present.

Making a Good Buffer

Acid buffer solutions that are centered should have pH values that are equal to the pK_a of the weak acid. As an example, using the weak acid HA and its conjugate A^-, the K_a expression for this acid is:

$$K_a = \frac{[H^+][A^-]}{[HA]}$$
which can be rearranged to
$$[H^+] = K_a \frac{[HA]}{[A^-]}$$

Taking the negative log of both sides of the equation on the right provides us with a simple relationship between the pH and the pK_a of an acid–base buffer system. This equation is called the Henderson–Hasselbalch equation:

$$pH = pK_a + \log\frac{[A^-]}{[HA]}$$

Because the acid and its conjugate appear in both the numerator and denominator, if they are of equal size, their concentrations will cancel. This tells us that the H_3O^+ concentration in such a mixture is equal to the K_a, or the pH = pK_a. In fact, the pH of a buffer is regulated by some combination of the K value of the weak electrolyte chosen and the ratio of conjugate to acid (or base). This is practical information, since it is unlikely that a particular weak acid or base will have exactly the right K value to create a buffer of a desired pH. However, by selecting an acid or base with a pK_a close to the desired pH and altering the ratio of acid (or base) to conjugate, a buffer with any desired pH can be made.

The amount of acid or base a buffer can absorb before a significant pH change is called its capacity. This is governed by the concentrations of the weak acid—or base—and the conjugate. While the pH is supposed to remain reasonably constant, it *does* change, even before one of the concentrations becomes exhausted. The determination of the pH of a buffer solution after the addition of acid or base is a relatively simple process. Using our hypothetical acid, HA, adding an amount of strong acid (x) will shift the equilibrium to the left:

$$HA \quad + \quad H_2O \quad \Leftrightarrow \quad H_3O^+ \quad + \quad A^-$$

	+x		x	−x
Eq	HA + x			$A^- - x$

Note that H_3O^+ readily reacts with A^-, increasing the amount of HA and decreasing the amount of A^-.

It is simpler to use moles in these calculations because the volume of the mixture will typically change because most added acids or bases would be in aqueous form. This volume change would necessitate a recalculation of the molarities of HA and A^-, except that the new volume will cancel in the K_a expression!

$$pH = pK_a + \log\frac{[A^- - x]}{[HA + x]}$$

From the expression above, it is clear that the change in the ratio of A^- to HA affects the pH. As long as x is small compared to the moles of A^- and HA, the change in pH will be small. The magnitude of x in relation to that amount of acid (or base) and conjugate is what determines the buffer capacity mentioned earlier.

A similar, but opposite, analysis applies to the case of adding strong base (x). HA will react with base in a neutralization reaction. The situation is:

$$HA \quad + \quad OH^- \quad \Leftrightarrow \quad H_2O \quad + \quad A^-$$

	−x	x		+x
Eq	HA − x			$A^- + x$

To calculate the pH of the solution after base is added the expression would be;

$$pH = pK_a + \log\frac{[A^- + x]}{[HA - x]}$$

Preparing a buffer solution of a particular pH would now seem to be a fairly simple process. After consulting a table of K_a values, an acid (or base) with a pK_a close to the desired pH is selected, and then the ratio of conjugate to acid (or base) is adjusted to achieve the desired pH value. When this is done in the lab [as in this experiment], the results are only approximate because the actual solution pH is also affected by both the solution temperature and the pH of the distilled water used to make the buffer. Neither of these factors is accounted for in our original buffer calculation.

The buffer system chosen for this experiment consists solely of ionic species. The equilibrium involved may be written as:

$$H_2PO_4^- + H_2O \Leftrightarrow H_3O^+ + HPO_4^{2-}$$

$$K_a = 2.3 \times 10^{-7}$$

Phosphoric acid (H_3PO_4) changes pretty quickly into dihydrogen phosphate, or $H_2PO_4^-$. This dihydrogen phosphate is an excellent buffer, since it can either accept a proton and re-form phosphoric acid, or it can give off another proton and become hydrogen phosphate, or HPO_4^{2-}. These ions (which are species derived from phosphoric acid) can be obtained as potassium and sodium salts. The value of K_a ($pK_a = 6.64$) suggests a reasonable target pH range for this buffer system of about 6.30 to 6.90.

In this experiment, you will be assigned a target pH for this system. You can calculate the ratio of buffer component concentrations by using the Henderson–Hasselbalch equation:

$$pH = pK_a + \log \frac{[A^-]}{[HA]}$$

Once the solution is prepared, the pH can be measured in the usual fashion. The capacity of the buffer can also be checked by calculating the volume of strong acid or base (0.50 M HCl or NaOH) needed to change the pH of a 25 mL sample by 0.50 pH units and then observing the result when this is done in the lab. A few additional tests on the buffer solution complete the experiment.

Finally, it should be said that in practice, laboratory chemists do not do these kinds of calculations before preparing buffer solutions. Buffer solutions with integer pH values are readily available commercially (someone else has done all of the calculations!). When buffers with noninteger pH values are required, approximate solutions are prepared, and strong acid or base is carefully added while the pH of the mixture is monitored with a calibrated pH electrode to achieve the desired pH.

Buffer Preparation: An Example

For today's lab, both components of the conjugate acid–base pair are to be weighed out separately to obtain the desired ratio and then dissolved in water.

Here's how to do it

Problem: Prepare 1.0 L of a 0.50 M potassium phosphate buffer at pH 6.90, assuming the availability of solid H_3PO_4 ($pK_a = 2.00$), KH_2PO_4 ($pK_a = 6.64$), K_2HPO_4 ($pK_a = 12.00$), and K_3PO_4.

Step 1. Determine the principal components of the buffer system. This is easy for a monoprotic system. For diprotic or polyprotic systems, this can vary depending on the desired pH. Identify the conjugate acid–base pair and write out the equilibrium.

In this case, the desired pH (6.90) is closest to the pK_a of the second ionization:

$$H_2PO_4^- + H_2O \Leftrightarrow H_3O^+ + HPO_4^{2-} \qquad pK_a = 6.64$$

(You might notice that the pK_a in your text is different from the one given here. This is because the pK_a given here takes into account the activity coefficient of the buffer. Please see A. A. Green, "The Preparation of Acetate and Phosphate Buffer Solutions of Known pH and Ionic Strength," *J. Am. Chem. Soc. 55*, 2331 [1933] for further explanation.)

Step 2. Calculate the desired ratio of the conjugate acid–base pair using the Henderson–Hasselbalch equation:

$$pH = pK_a + \log \frac{[HPO_4^{2-}]}{[H_2PO_4^-]}$$

$$\frac{[HPO_4^{2-}]}{[H_2PO_4^-]} = 10^{pH - pK_a}$$

In this case,

$$\frac{[HPO_4^{2-}]}{[H_2PO_4^-]} = 10^{6.90 - 6.64} = 10^{0.26} = 1.8$$

We can solve it by using the two statements of what we know about the buffer acid and base concentrations:

$$[HPO_4^{2-}] = 1.8\,[H_2PO_4^-]$$

and

$$[HPO_4^{2-}] = 0.50\ M - [H_2PO_4^-]$$

If we set the two definitions of $[HPO_4^{2-}]$ equal to each other and solve for $[H_2PO_4^-]$, we get:

$$1.8\ [H_2PO_4^-] = 0.50\ M - [H_2PO_4^-]$$

$$1.8\ [H_2PO_4^-] + [H_2PO_4^-] = 0.50\ M$$

$$2.8\ [H_2PO_4^-] = 0.50\ M$$

$$[H_2PO_4^-] = 0.50\ M/2.8 = 0.18\ M$$

We can then substitute the $[H_2PO_4^-]$ value back into one of the original two equations to solve for $[HPO_4^{2-}]$:

$$[HPO_4^{2-}] = 0.50\ M - 0.18\ M = 0.32\ M$$

Step 3. Determine the most feasible means of obtaining the desired components. In this case, the obvious choice is to weigh out the desired amounts of the potassium salts (KH_2PO_4 and K_2HPO_4), which upon dissolution will completely ionize, giving both components of the acid—base pair.

Step 4. Calculate the required amount of each material. Multiplying by 1 L, we know that we need 0.18 moles of KH_2PO_4 and 0.32 moles of K_2HPO_4. After calculating the formula weights of these, we can obtain the required masses:

KH_2PO_4: $(0.18\ mol)(136.1\ g/mol) = 25\ g\ KH_2PO_4$

K_2HPO_4: $(0.32\ mol)(174.2\ g/mol) = 56\ g\ K_2HPO_4$

Step 5. Prepare the buffer. Weigh out 25 g of KH_2PO_4 and 56 g of K_2HPO_4, and dissolve in ~900 mL of distilled water. Check the pH and adjust if necessary. Bring the total volume to 1 L.

Procedure

SAFETY NOTES: Mono- and di-basic phosphate salts are caustic and can cause skin irritation and burns, so take all due precaution when handling.

GENERAL INSTRUCTIONS: Each student will prepare and test his or her own buffer. If interested, you can perform an acid and alkaline self-test by collecting a small piece of litmus paper and applying some saliva to the paper; this will indicate whether your body fluids are too acidic or too alkaline. Litmus paper will change color to indicate pH levels of body fluids. These tests should be performed before eating or an hour after eating.

Part I: Comparing the Behavior of the Buffer with Distilled Water

Obtain a pH meter. Calibrate the pH meter for data collection.

Place 25 mL of distilled, deionized (DI) water in a clean, dry 100 mL beaker. Measure the pH of the DI water and record the pH in your lab notebook.

Using a disposable pipette, add 1 drop (~0.05 mL) of 0.50 M HCl to the beaker. Record the change in pH.

Discard the solution down the sink with lots of water and place a fresh 25 mL of distilled, DI water in a clean, dry 100 mL beaker.

Using a disposable pipette, add 1 drop (~0.05 mL) of 0.50 M NaOH to the beaker. Record the change in pH.

Part II: Preparing and Testing the Buffer Solution for the Assigned pH

Using the calculations from your prelab work, prepare 100 mL of your buffer at the assigned pH.

Once complete, measure out a 25 mL aliquot of the prepared buffer and place it in a clean, dry 100 mL beaker. Measure and record the pH.

Part III: Testing the Buffer Capacity with HCl and NaOH

Using the 25 mL of the buffer above, add 1 mL of 0.5 M HCl solution to the beaker and mix well. Record the new pH. Repeat the addition of 1 mL of 0.5 M HCl and continue to record the pH until the pH has dropped by 0.5 pH units.

Discard the used buffer solution down the sink with lots of water and place a fresh 25 mL of your prepared buffer in a clean, dry 100 mL beaker.

Using the 25 mL of buffer above, add 1 mL of 0.5 M NaOH solution to the beaker and mix well. Record the new pH. Repeat the addition of 1 mL of 0.5 M NaOH and continue to record the pH until the pH has risen by 0.5 pH units.

Part IV: Determining the Effect on the pH of Diluting the Buffer

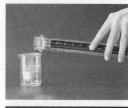

Discard the used buffer solution down the sink with lots of water and place a fresh 12.5 mL of your prepared buffer in a clean, dry 100 mL beaker. Add 12.5 mL of distilled water to the beaker to dilute the buffer.

Repeat all of Part III (using HCl, and then NaOH, with a fresh sample of diluted buffer) to determine the buffer capacity of the diluted buffer.

Report Contents and Questions

The purpose should be several well-constructed sentences describing what your experiment was designed to accomplish and the criteria used to determine success. These sentences should include both concepts and techniques. The procedure section should reference the lab manual and include any changes made to the procedure during the lab.

The data section should include tables with the following data for distilled water, your buffer, and your diluted buffer: measured pH, calculated concentration of HA and A⁻ after each addition of 1.0 mL of HCl or NaOH, and calculated pH after each addition.

A sample calculation of how you prepared the buffer solution at the assigned pH and a sample calculation for determining the acid and base concentration after the addition of HCl or NaOH should be included.

The conclusion section should include several paragraphs with a summary statement regarding the buffer and pH used, the buffer capacity, and the comparison of the calculated pH values and the experimental pH values recorded in lab. A discussion on the similarities and differences between the buffer and dilute buffer should be included. Any possible errors should be discussed as usual.

Answer the following question:

1) Human blood contains a buffering system. What are the key molecules in this buffer, and why is it necessary?

Experiment 21
Laboratory Preparation

Name: _____ Date: _____

Instructor: _____ Sec. #: _____

Show all work for full credit.

1) Read the background, procedure, and report sections of the lab experiment carefully and develop a hypothesis of what information you expect to gain from the completion of the lab experiment.

2) Create any and all tables you might need to collect data during the experiment and then transfer those tables into your lab manual for use during the lab.

Experiment 21

Prelaboratory Assignment

Name: _____ Date: _____

Instructor: _____ Sec. #: _____

Show all work for full credit.

1) You need to prepare 1.00 L (in a volumetric flask) of 0.50 M phosphate buffer, pH 6.21. Use the Henderson–Hasselbalch equation with a value of 6.64 for pK_2 to calculate the quantities of K_2HPO_4 and KH_2PO_4 you need to add to the flask. Record the steps of these calculations and use them as a guide for the calculation you will have to make in the laboratory. (Calculate the intermediate values to at least one more significant figure than required.)

What is the $[K_2HPO_4]$ to $[KH_2PO_4]$ ratio you will need?

What concentrations of $[K_2HPO_4]$ and $[KH_2PO_4]$ will you need to make the total concentration 0.50 M? (Units required.)

How many moles of K_2HPO_4 and KH_2PO_4 will you need? (Units required.)

What mass of K_2HPO_4 and KH_2PO_4 will you need? (Units required.)
(FW of K_2HPO_4 = 174.2 g/mol. FW of KH_2PO_4 = 136.1 g/mol.)

2) Buffer capacity refers to the amount of acid or base a buffer can absorb without a significant pH change. It is governed by the concentrations of the conjugate acid and base forms of the buffer. A 0.5 M buffer will require five times as much acid or base as a 0.1 M buffer for a given pH change. In this problem, you begin with a buffer of known pH and concentration and calculate the new pH after a particular quantity of acid or base is added. In the laboratory, you will carry out some stepwise additions of acid or base and measure the resulting pH values.

Starting with 60 mL of 0.50 M phosphate buffer, pH = 6.83, you add 1.7 mL of 1.00 M HCl. Using the Henderson–Hasselbalch equation with a pK_2 for phosphate of 6.64, calculate the following values to complete the ICE table.

What is the composition of the buffer to begin with, both in terms of the concentration and the molar quantity of the two major phosphate species? (Units required.)

What is the molar quantity of H_3O^+ added as HCl, and the final molar quantity of HPO_4^{2-} and $H_2PO_4^-$ at equilibrium?

What is the new $HPO_4^{2-}/H_2PO_4^-$ ratio, and the new pH of the solution? (NOTE: You can use the molar ratio rather than the concentration ratio because both species are in the same volume.)

Now you take another 60 mL of the 0.50 M pH 6.83 buffer and add 3.7 mL of 1.00 M NaOH. Using steps similar to those above, calculate the new pH of the solution.

Experiment 22
Entropy, Free Energy, and Chemical Equilibrium

Introduction

When initially learning the basics of general chemistry, solubility was discussed in a general sense with respect to being an all or nothing physical property of ionic compounds. The initial solubility rules stated which salts were soluble in water and which were not. As with many concepts introduced in the first term, that is not the whole story. Now that you have learned more about the concepts of equilibrium and intermolecular forces, we can discuss solubility in more detail.

The solubility of a substance is actually dependent on the forces holding the crystal together (the lattice energy) and on the solvent acting on the crystal. For now, we will consider only water as the solvent. As a solid dissolves, the ions in solution are surrounded by water molecules via a process called hydration. During hydration, energy is released. The extent to which the energy of hydration exceeds the lattice energy determines the solubility. Thus, solubility is not all or nothing; rather, it is an equilibrium process that is dependent on the energy availability of the system. Almost all substances are soluble to some degree.

Le Chatelier's principle allows us to predict the effects of changes in temperature, pressure, and concentration on a system at equilibrium. It states that if a system at equilibrium experiences a change, its equilibrium will shift to compensate for the change. This is why solubility is affected by changes in temperature. For every reversible reaction, one direction is endothermic, and the other is exothermic. A reaction is endothermic if it absorbs heat from its surroundings and exothermic if it releases heat to the surroundings. If you increase the temperature, then the endothermic reaction will be favored because it will take in some of the excess heat and use it to force the reaction in the forward direction. If you decrease the temperature, the exothermic reaction will be favored because it will replace the heat that was lost.

The purpose of this experiment is to investigate the solubility of $Ca(OH)_2$ at different temperatures. You will also be asked to determine the K_{sp}, ΔG^o, ΔH^o, and ΔS^o for the $Ca(OH)_2$ solution at each temperature. By creating saturated solutions over a range of temperatures and titrating those solutions with a standardized hydrochloric acid solution, the concentration of $Ca(OH)_2$ can be determined. At the conclusion of the experiment, you should be able to discuss the thermodynamics of $Ca(OH)_2$ solubility with respect to Le Chatelier's principle.

Background

The solubility of a compound is defined as the amount of solute that can be dissolved in a particular solvent, normally water. Until now, you have most likely treated solubility as a "yes" or "no" property of a substance. In other words, after memorizing the solubility rules, you know that salts of nitrates, nitrites, chlorates, bicarbonates, and acetates are soluble, while the salts of carbonates, phosphates, and chromates are generally insoluble (Group I and ammonium salts excluded). Well, if the truth be told, ALL ionic compounds are soluble in water to some varying degree.

Solubility

If we consider solubility in terms of a reaction where a solid ionic reagent dissolves into its component ions, then the reaction is governed by the same rules of equilibrium that affect all other chemical reactions. And, since dissolving is an equilibrium process, it seems logical that it can be described by an equilibrium constant, referred to as the solubility product constant (K_{sp}). The K_{sp} describes the equilibrium between the soluble and insoluble portion of the solute. The solubility constant expression is derived by following the same rules you would use when writing any other equilibrium expression.

$$\text{In general, } A + B \Leftrightarrow C + D \quad \text{gives an expression } K = \frac{[C][D]}{[A][B]}$$

When discussing solubility, it is important to understand the factors affecting the process. In particular, the degree of an ionic compound's solubility is dependent upon several factors, including (1) the intermolecular forces between the solute and solvent; (2) the change in entropy accompanying the process; (3) the concentration of the products and the reactants; and (4) the factor that you will be directly observing in this experiment, the temperature.

Calcium Hydroxide

The ionic compound that you will be dealing with in this experiment is unique. For most salts, the solubility of the compound is greater at higher temperatures than lower temperatures. However, the solubility of calcium hydroxide is greatest at lower temperatures, a behavior completely opposite of what is generally expected. To observe the effect of temperature on solubility and equilibrium, various solutions of calcium hydroxide will be produced at different temperatures. Once equilibrium has been established at each temperature, any excess solid will be removed by filtration; then the filtrate will be titrated with standard HCl to establish the concentration of hydroxide ion.

To further investigate some of the principles just discussed, let's look at some sample data obtained from a similar experiment involving $Ca(OH)_2$. Let's assume that after titrating 35.00 mL of the solution prepared at 25 °C, we found that we had to add 15.10 mL of 0.050 mol/L HCl to

271

reach the end point. Recalling that the end point occurs when the number of moles of acid equals the number of moles of base, we can easily figure out the concentration of the hydroxide ion in this particular solution by multiplying the volume of acid added by its respective concentration:

$$\text{moles of acid} = 0.01510 \text{ L} \times 0.050 \text{ } \tfrac{mol}{L} = 7.6 \times 10^{-4} \text{ moles}$$

Setting the number of moles of acid equal to the number of moles of base, and then dividing by the volume of $Ca(OH)_2$, we arrive at the concentration of the hydroxide ion:

$$\frac{7.6 \times 10^{-4} \text{ moles}}{0.035 \text{ L}} = 0.022 \text{ M}$$

The equilibrium reaction for dissolving $Ca(OH)_2$ is shown below:

$$Ca(OH)_2(s) \Leftrightarrow Ca^{2+}(aq) + 2OH^-(aq)$$
$$K_{sp} = [Ca^{2+}][OH^-]^2$$

Since the equilibrium reaction shows a 2:1 molar ratio of OH^- to Ca^{2+}, we can easily arrive at the concentration of the calcium ion by dividing the concentration of the hydroxide ion by 2 giving a calcium ion concentration $[Ca^{2+}]$ of approximately 0.011 mol/L. Taking these values and substituting into the K_{sp} expression, we arrive at a value of about 5.3×10^{-6} for the solubility product constant of $Ca(OH)_2$ at 25 °C. With the same process being carried out at various temperatures, you will be able to see the temperature dependence of solubility.

Thermodynamics of Solubility

Remembering that K_{sp} is a special form of an equilibrium constant, it seems logical that other thermodynamic factors such as enthalpy (ΔH), entropy (ΔS), and Gibbs free energy (ΔG) will also be dependent upon temperature changes. In fact, the equilibrium constant K_{sp} is directly correlated with the Gibbs free energy of an ionic compound's dissociation, as shown in the following equation where R is the ideal gas constant (8.3145 J/mol K) and T is the experimental temperature in Kelvin.

$$\Delta G = -RT \ln(K_{sp})$$

For our example run at 25 °C, 298 K, the value of ΔG comes out to be about 30. kJ/mol.

After we have calculated ΔG at the various temperatures, we can generate a plot of ΔG versus temperature, in Kelvin, obtain a slope and a y-intercept, and then use the following relationship to calculate ΔH and ΔS for the dissociation process

$$\Delta G = \Delta H - T\Delta S$$

Note that this equation is in the familiar $y = mx + b$ format, with the slope and y-intercept equal to $-\Delta S$ and ΔH, respectively. At this point, it will be your job to figure out the significance of these two parameters!

Procedure

SAFETY NOTES: $Ca(OH)_2$ is caustic, and HCl is corrosive. All due precautions should be taken in their handling.

GENERAL INSTRUCTIONS: Students should work in groups and share data for this experiment so that all four titrations can be completed in the time allotted, that is, each student should complete two of the four different temperature titrations. Be sure to dispose of your calcium hydroxide solutions as directed by your instructor.

Part I: Titration of a Room Temperature, Saturated Ca(OH)₂ Solution

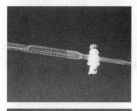

Obtain a buret and rinse it well with DI water.

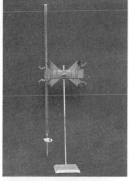

Use a clamp to attach the buret to the ring stand on your bench and then set a magnetic stirrer (hotplate) underneath it.

Pour ~30 mL of 0.05 M HCl solution into a clean, dry 50 or 100 mL beaker. Record the concentration of the acid.

Rinse the inside of the buret with a few milliliters of the HCl solution, making sure all the walls of the buret are coated. Collect the rinse solution in a beaker labeled "Waste."

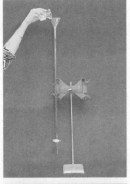

With the stopcock closed, pour the remainder of the HCl solution into the buret. Place a small waste beaker under the buret and slowly open the stopcock until the solution begins to drip out.

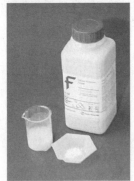

Close the stopcock and make sure there are no air bubbles in the tip of the buret. Also make sure that the level of liquid in the buret is at or below the 0.00 mL line. Dispose of the wash solution in your waste beaker.

Add ~100 mL of distilled water to a 250 mL Beaker and, using a thermometer, determine and record its temperature to 0.1 °C. Add 2.0 g of solid $Ca(OH)_2$ with constant stirring until the salt no longer dissolves.

Using a funnel, strain this solution through qualitative filter paper into a clean, dry 125 mL Erlenmeyer filter flask. NOTE: Only filter small amounts of solution at a time to avoid "swamping" the filter paper. The filtrate solution should be clear when filtration is complete. A cloudy filtrate indicates the presence of particulates that will throw off the titration results. Refilter any cloudy filtrates until they are clear.

Pipet 10.00 mL of this *clear* solution into another clean, dry 100 mL Beaker. Add ~25 mL of distilled water and 10 to 12 drops of the bromothymol blue indicator. Gently slide a magnetic stir bar into the flask.

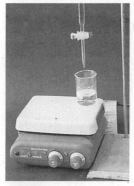

Set the flask on the stirrer and begin stirring at a gentle rate. (Make sure there is no splashing of the solution.) Record to two decimal places the beginning volume level of HCl in the buret.

Begin the titration, allowing the titrant to fall from the buret at a rapid drop-by-drop pace. During this addition, the initial blue color will begin to turn green and then yellow. The end point is reached when the entire solution turns yellow. Your goal is to deliver the exact volume needed to reach the end point. When you have reached the end point, record the final volume in the buret (again to two decimal places) and calculate the volume of titrant delivered.

Pour the titrated solution into the waste container, but don't lose your stir bar.

Part II: Titration of an Elevated Temperature, Saturated Ca(OH)$_2$ Solution

Add about 100 mL of distilled water to a 250 mL beaker. Place it on a hotplate and bring the water to between 50 °C and 70 °C. After the water has been at temperature for several minutes, add ~2 g of solid Ca(OH)$_2$ to the water with constant stirring until the salt no longer dissolves. Take a final temperature reading at saturation.

Using a funnel, *quickly* strain this solution through qualitative filter paper into a clean, dry 125 mL Erlenmeyer filter flask. NOTE: Only filter small amounts of solution at a time to avoid swamping the filter paper. The filtrate solution should be clear when filtration is complete. A cloudy filtrate indicates the presence of particulates that will throw off the titration results. Refilter any cloudy filtrates until they are clear.

Pipet 10.00 mL of this *clear* solution into another clean, dry 100 mL Beaker. Add approximately 25 mL of distilled water and 10 to 12 drops of the bromothymol blue indicator. Gently slide a magnetic stir bar into the flask.

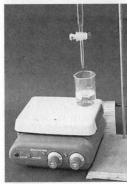

Set the flask on the stirrer and begin stirring at a gentle rate. (Make sure there is no splashing of the solution.) Record to two decimal places the beginning volume level of HCl in the buret.

Begin the titration, allowing the titrant to fall from the buret at a rapid drop-by-drop pace. During this addition, the initial blue color will begin to turn green and then yellow. The end point is reached when the entire solution turns yellow. Your goal is to deliver the exact volume needed to reach the end point. When you have reached the end point, record the final volume in the buret (again to two decimal places) and calculate the volume of titrant delivered.

Pour the titrated solution into the waste container, but don't lose your stir bar.

Part III: Titration of a Low-Temperature, Saturated $Ca(OH)_2$ Solution

Add ~100 mL of distilled water to a 250 mL beaker. Place it in a salted ice bath for at least 5 min. Add solid $Ca(OH)_2$ with constant stirring until the salt no longer dissolves.

Remove the flask from the ice bath and measure the temperature of the cold solution.

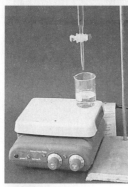

Using a funnel, *quickly* strain this solution through qualitative filter paper into a clean, dry 125 mL Erlenmeyer filter flask. NOTE: Only filter small amounts of solution at a time to avoid swamping the filter paper. The filtrate solution should be clear when filtration is complete. A cloudy filtrate indicates the presence of particulates that will throw off the titration results. Refilter any cloudy filtrates until they are clear.

Pipet 10.00 mL of this *clear* solution into another clean, dry 100 mL Beaker. Add ~25 mL of distilled water and 10 to 12 drops of the bromothymol blue indicator. Gently slide a magnetic stir bar into the flask.

Set the flask on the stirrer and begin stirring at a gentle rate. (Make sure there is no splashing of the solution.) Record to two decimal places the beginning volume level of HCl in the buret.

Begin the titration, allowing the titrant to fall from the buret at a rapid drop-by-drop pace. During this addition, the initial blue color will begin to turn green and then yellow. The end point is reached when the entire solution turns yellow. Your goal is to deliver the exact volume needed to reach the end point. When you have reached the end point, record the final volume in the buret (again to two decimal places) and calculate the volume of titrant delivered.

Pour the titrated solution into the waste container, but don't lose your stir bar.

Part IV: Titration of a ? Temperature, Saturated Ca(OH)₂ Solution

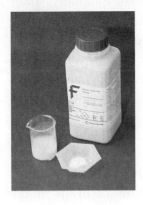

Based on your calculation of the molar solubility of the Ca(OH)₂ solution at low, room, and high temperature, choose another temperature at which to run the titration. Follow the same procedure as in Parts I through III.

Report Contents and Questions

The purpose should be several well-constructed sentences describing what your experiment was designed to accomplish and the criteria used to determine success. These sentences should include both concepts and techniques. The procedure section should reference the lab manual and include any changes made to the procedure during the lab. The different temperatures that you ran should be specified. Also include the name of your partner and clearly identify the data contributed by each person.

The data section must include a balanced equation of the reaction being studied. A table should be made with the following for each temperature studied; **(a)** the average volume of standard HCl in each trial ($V[\text{mL}] = V_{final} - V_{inital}$) A second table should be provided that summarizes all of the calculated information for each temperature: **(a)** the hydroxide concentration [OH⁻], **(b)** the molar solubility, **(c)** the K_{sp}, and **(d)** ΔG should be included.

Determine the hydroxide concentration as follows. Use the concentration of standard HCl (this should have been recorded in lab) and the V to find the moles of HCl. From the moles of HCl and the balanced equation, determine the moles of OH⁻. Be sure to show the mole ratio used. From the moles of OH⁻ and volume that you titrated out (i.e., the volume of filtrate you put in your Erlenmeyer flask), you can calculate [OH⁻].

Once you have [OH⁻] × you can find [Ca²⁺] using the stoichiometry of the solubility equation.

The K_{sp} at each temperature is determined by:

$$Ca(OH)_2(s) \Leftrightarrow Ca^{2+}(aq) + 2OH^-(aq)$$
$$K_{sp} = [Ca^{2+}][OH^-]^2$$

NOTE: The concentration of Ca²⁺ is equal to the molar solubility. List the molar solubility of the compound for each temperature.

ΔG for each temperature can be determined using $\Delta G = -RT \ln K_{sp}$, where R is the gas constant, T is temperature in Kelvin, K_{sp} is the equilibrium constant, and ΔG is Gibbs free energy. Make sure you have the correct units. ΔH and ΔS for this reaction should also be stated and are determined from a graph (see calculations).

The calculation section should include a graph of ΔG versus temperature (in Kelvin) AND a graph of molar solubility versus temperature. All graphing guidelines apply. Example calculations of the following should also be included: $[OH^-]$, molar solubility, K_{sp}, ΔG, ΔH, and ΔS. The values for ΔH and ΔS are determined from the free energy graph where the slope of the graph is $-\Delta S$ and the y-intercept of the graph is ΔH. Make sure that the units are correct in all of your calculations and answers.

The conclusion section should be in paragraph form and include $[OH^-]$ for each temperature and the molar solubility of $Ca(OH)_2$ at each temperature. The trend seen on your molar solubility versus temperature graph should be explained. Another discussion of the values, trends, and discrepancies in the values for K_{sp}, ΔG, ΔH, and ΔS should be included. Finally, be sure to discuss any possible experimental errors.

Answer the following questions:

1) Why do you have to filter the $Ca(OH)_2$ solution before titrating it?

2) Why don't you have to accurately record the amount of $Ca(OH)_2(s)$ used in the saturated solution?

Experiment 22
Laboratory Preparation

Name: _____ **Date:** _____

Instructor: _____ **Sec. #:** _____

Show all work for full credit.

1) Read the background, procedure, and report sections of the lab experiment carefully and develop a hypothesis of what information you expect to gain from the completion of the lab experiment.

2) Create any and all tables you might need to collect data during the experiment and then transfer those tables into your lab manual for use during the lab.

Experiment 22
Prelaboratory Assignment

Name: _____ Date: _____

Instructor: _____ Sec. #: _____

Show all work for full credit.

1) You add an excess of $Ca(OH)_2$ to water maintained at a particular temperature, stir until the solution is saturated, filter, then determine the [OH⁻] in the solution by titration with acid. Titration of 10.0 mL of the calcium hydroxide solution to the end point requires 4.86 mL of 0.070 M HCl solution.

   What is the molar quantity of OH⁻ in the 10.0 mL of solution?

   What are the concentrations of Ca^{2+} and OH⁻?

   What is the solubility of $Ca(OH)_2$ under these conditions?

   What is the K_{sp} for $Ca(OH)_2$ under these conditions?

2) You measure the solubility of a salt at four different temperatures and calculate the following K_{sp} values:

Temperature (°C):	15	31	51	70
K_{sp}:	2.89×10^{-2}	4.11×10^{-2}	6.08×10^{-2}	8.45×10^{-2}

   ΔG^o_{soln} for the solubility process is given by the relationship $\Delta G^o_{soln} = -RT \ln K_{sp}$ where $R = 8.314$ J/(K × mol), and T is the absolute temperature in degrees K.

   Calculate ΔG^o_{soln} (in kJ/mol) at each of these temperatures. As the temperature increases, how does the solubility of the salt change? According to Le Chatelier's principle, dissolving this salt should therefore be which type of process? In other words, what is the sign of ΔH^o_{soln}?

3) It is possible to determine both $\Delta H^{o}{}_{soln}$ and $\Delta S^{o}{}_{soln}$ for the solution process from this data. One or both of two graphical methods can be used:

Method 1: From the equation:
$$\Delta G^{o}{}_{soln} = \Delta H^{o}{}_{soln} - T\Delta S^{o}{}_{soln}$$

a plot of $\Delta G^{o}{}_{soln}$ versus T will give $-\Delta S^{o}{}_{soln}$ as the slope and $\Delta H^{o}{}_{soln}$ as the intercept.

Method 2: Alternatively, rearrange the equation relating K_{sp} and $\Delta G^{o}{}_{soln}$:

$$\ln K_{sp} = -\frac{\Delta G^{o}{}_{soln}}{RT} = -\frac{\Delta H^{o}{}_{soln}}{RT} + \frac{\Delta S^{o}{}_{soln}}{R}$$

and plotting $\ln K_{sp}$ versus $\frac{1}{T}$ will give $-\frac{\Delta H^{o}{}_{soln}}{RT}$ as the slope and $\frac{\Delta S^{o}{}_{soln}}{R}$ as the intercept.

Using one or both of these graphical methods, determine $\Delta H^{\circ}{}_{soln}$ (in kJ/mol) and $\Delta S^{\circ}{}_{soln}$ (in J/(K $\times$ mol)) for this salt. (Be sure to play close attention to your units and signs in both your plots and your calculations.)

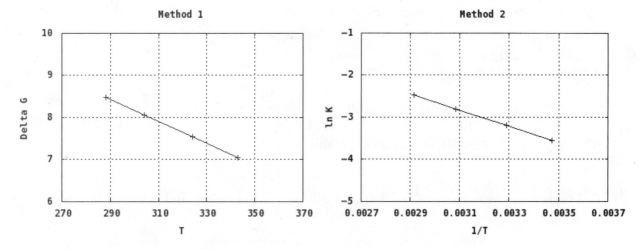

Experiment 23
Electrochemistry: The Nernst Equation

Introduction

This experiment will expand on the investigation of electrochemistry and its simple redox reactions by building a mock voltaic cell and measuring it's potentials at room temperature along with various temperatures. Electrochemistry is important in understanding the development of batteries and other energy storage devices, but another reason for studying electrochemistry is a more personal reason, being the human body and how it runs on electrochemical reactions.

Adenosine triphosphate (ATP) is the energy molecule used by the body to fuel cellular reactions. The chemical processes used to produce ATP are oxidation–reduction reactions and are known as the electron transport chain.

Adenosine triphosphate (ATP)

As standard state is defined as 1.00 atmosphere of pressure, 1.0 M, and 25 °C; biological systems are not at standard state. Many of you have plans to be medical doctors or work in the allied health field and will need to understand this process in great detail because many metabolic diseases are related to the improperly functioning enzymes in the electron transport chain. Thus, our expansion of this experiment to nonstandard temperatures is very important. In this experiment, we will determine the potentials of 10 cells at several temperatures and use a version of the Nernst equation to calculate the free energy, enthalpy, and entropy of those cells.

Background

In the later years of the eighteenth century, Luigi Galvani discovered that the nerves of frogs were capable of harboring electrical activity. Shocked by his discovery, Galvani rationalized that such activity could be found only in living tissues. A few years later, a man by the name of Alessandro Volta scientifically refuted Galvani's claim when he observed that electricity could also be produced through inorganic means. By using small metal sheets of copper and zinc separated by pieces of cloth soaked in acidic solution, Volta constructed what is believed to be the first apparatus capable of producing electricity. Through centuries of research and technological advances, electricity is now a crucial part of our everyday lives, an aspect so vital that it has its own chemical course of study—electrochemistry.

Electrochemistry

Electrochemistry is the study of the conversion between electrical and chemical energy. Reactions of this sort are also called redox reactions because they utilize two unique processes, oxidation and reduction, to transfer electrons. Oxidation involves the loss of one or more electrons from a chemical species, while reduction refers to the gain of electrons by a chemical species. These two definitions are easily remembered by the mnemonic OIL RIG (oxidation is loss, reduction is gain) Or LEO GER (lose an electron, oxidation; gain an electron, reduction).

When an oxidation and a reduction reaction are paired together in a redox reaction, electrons can flow from the oxidized species to the reduced species. For this reason, the oxidized species can also be referred to as the reducing agent or the reductant, while the reduced species is also known as the oxidizing agent or oxidant. The flow of electrons can be observed when an electrochemical cell is constructed. Two types of cells are (1) a galvanic cell, (also known as a voltaic cell), where a spontaneous reaction drives the electron flow and (2) an electrolytic cell, where an outside source of current is imposed on the cell to drive a chemical reaction.

Electrochemical Cells

When discussing an electrochemical cell, there are several new terms: *standard reduction potential* ($E°_{red}$); *half-cell*; *electrode*; *anode*; and *cathode*. The first and most important concept here is standard reduction potential ($E°_{red}$), which is a numerical value of a half-reaction for the reduction of a particular metal. The standard reduction potential is a quantitative measure (in volts) of a particular substance's tendency to accept electrons under the standard conditions of 1.0 atm, 298 K and 1.0 M. A table of standard reduction potentials is in Appendix D in your textbook.

Observing the table, you will notice that the series of half-reactions is listed in descending order from the most positive value to the most negative value. In general, the more positive the standard reduction potential is for a particular substance, the stronger its tendency is to be reduced, or to gain electrons. The standard oxidation potential ($E°_{ox}$), is equal in value, but opposite in sign, to the standard reduction potential; the value of $E°_{ox}$ measures the tendency of a particular substance to lose electrons, or become oxidized. For example, the dichromate ion $(Cr_2O_7{}^{2-})$ has a standard reduction potential of $+1.33$ V, so under standard conditions, the ion will have a standard oxidation potential of -1.33 V.

The remaining new concepts deal entirely with electrochemical cells and are critical components of their construction. The system is composed of two parts commonly referred to as half-cells. A half-cell is the part of the electrochemical cell that houses the electrode. The electrode, as described by Faraday, can be either an anode, the electrode where oxidation occurs, or a cathode, the electrode where reduction takes place. The metal with the greater (more positive) potential will be the anode. For a clearer picture of what we have just described, please carefully note the figure below:

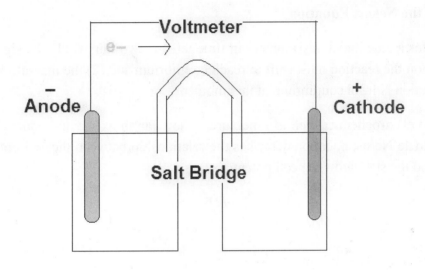

Calculating Cell Potentials

A concept not mentioned in the previous section is overall cell potential ($E°_{cell}$), which in brief is simply the sum of the standard reduction potential ($E°_{red}$) and the standard oxidation potential ($E°_{ox}$). Mathematically, the overall cell potential is expressed as:

$$E°_{cell} = E°_{red} + E°_{ox}$$

Using our diagram above, let's make a reasonable hypothesis as to what we believe the overall potential will be for a cell formed between zinc and copper metals. We will use a 1.0 M solution of zinc nitrate and zinc metal as the anode (left) and copper nitrate and copper metal as the cathode (right). Using the table of the half reaction reduction potentials in Appendix D, we find:

$$Cu^{2+}(aq) + 2e^- \Leftrightarrow Cu^0(s) \qquad E°_{red} = +0.34 \text{ V}$$

$$Zn^0(s) \Leftrightarrow Zn^{2+}(aq) + 2e^- \qquad E^\circ_{ox} = +0.76 \text{ V}$$

Using the equation above for the overall cell potential, we can arrive at the estimated value:

$$E^\circ_{cell} = E^\circ_{red} + E^\circ_{ox} = +0.34 \text{ V} + (+0.76 \text{ V}) = +1.10 \text{ V}$$

The actual reading on the voltmeter in the diagram above should confirm the value we have just calculated. However, when you construct this cell in the lab, the voltmeter reading probably will not be exactly +1.10 V. Remember, you are not working at standard conditions. However, if the ratio of concentrations of the corresponding salts is 1:1, any differences in the readings can be explained by other sources of error such as resistance in the salt bridge or lack of calibration of the voltmeter.

Free Energy and the Nernst Equation

The value of ΔG for a reaction at any moment in time tells us two things: (1) the sign of ΔG tells us in which direction the reaction must shift to reach equilibrium and (2) the magnitude of ΔG tells us how far the reaction is from equilibrium at that moment.

The potential of an electrochemical cell is a measure of how far an oxidation–reduction reaction is from equilibrium. The Nernst equation describes the relationship between the cell potential at any moment in time and the standard-state cell potential.

$$E = E^\circ - \frac{RT}{n\mathcal{F}} \ln Q$$

Let's rearrange this equation as follows,

$$n\mathcal{F}E = n\mathcal{F}E^\circ - RT \ln Q$$

We can now compare it with the equation used to describe the relationship between the free energy of reaction at any moment in time and the standard-state free energy of reaction.

$$\Delta G = \Delta G^\circ + RT \ln Q$$

These equations are similar because the Nernst equation is a special case of the more general free energy relationship. We can convert one of these equations to the other by taking advantage of the following relationships between the free energy of a reaction and the cell potential of the reaction when it is run as an electrochemical cell.

$$\Delta G = -n\mathcal{F}E \text{ and } \Delta G^\circ = -n\mathcal{F}E^\circ$$

In the second portion of this experiment, you will measure the potential of your galvanic cell at various temperatures, data that will give far more information about a particular reaction that you may be studying. As with any other type of chemical reaction, redox reactions are subject to changes in equilibrium brought about by changes in the overall temperature of the system. However, before we obtain the particular equilibrium constant (K) for our reaction, we need to calculate the change in free energy (ΔG). The free energy (ΔG) of a particular reaction is a measure of the spontaneity of the process and is defined mathematically by the equation we just derived:

$$\Delta G = -n\mathscr{F}E$$

In this equation, the variables are as follows: n is the number of moles of electrons transferred, $\mathscr{F}$ is Faraday's constant, and E is the observed overall cell potential. As seen above with copper, the setup is as follows:

$$\Delta G = -n\mathscr{F}E = -(2 \text{ mole}^-)(96.5 \tfrac{kJ}{V \cdot mol})(1.10 \text{ V}) = -212 \text{ kJ}$$

In this particular reaction, the ΔG is ~ -212 kJ, meaning that our example reaction is spontaneous when both products and reactants are in their standard states.

Now, with an approximate value for ΔG known, we can calculate the reaction's equilibrium constant from the relationship shown below.

$$K = e^{-\frac{\Delta G}{RT}}$$

For this equation, the value of ΔG is calculated as above, R is the ideal gas constant with units of J-mol^{-1}-K^{-1}, and T is the temperature in Kelvin. Realizing we performed our experiment at room temperature (~ 27 °C), we can quickly calculate the corresponding equilibrium constant.

$$K = e^{\frac{-(-212,000 \text{ J})}{8.3145\frac{J}{mol \cdot K}(300 \text{ K})}} = 8.2 \times 10^{36}$$

Thus, for our particular example run at room temperature, we get an approximate value of $K = 8.2 \times 10^{36}$! In other words, this reaction is driven completely to product at room temperature.

The second part of this experiment is aimed at furthering your understanding of how temperature can affect the free energy of a reaction as well as its equilibrium. By constructing a graph of ΔG versus temperature, entropy and enthalpy are easily derived from the slope and intercept of the resulting line, since the equation $\Delta G = \Delta H - T\Delta S$ is of the form $y = mx + b$.

Procedure

GENERAL INSTRUCTIONS: Because there are five metals, there are 10 possible cells you can create. Your instructor will assign each student one of the possible cells to build and test. Data from each student will then be pooled to allow you to build a potential series of all the metals.

Part I: Building the Cells

The cells will be constructed out of two large test tubes and a salt bridge created by the student.

Each bench can work as a group to make the agar–agar, as only a small milliliter amount is needed for each U-tube.

Place 50 mL of 0.1 M KNO_3 in a 250 mL beaker and, using a hot plate, bring it to a boil.
Remove the solution from the heat and add 0.50 g of agar–agar with constant stirring.

Obtain a U-tube.
Use a plastic pipet to fill the agar–KNO_3 gel into the inverted U-tube until the gel fills all but 0.5 in. of the tubes. Allow the gel to cool and solidify.

Soak half of some cotton plugs in each of the solutions you are using in your cell.
Fill the 0.5 in. spaces in the tubes with cotton plugs, soaked side in. Make sure that the cotton contacts the agar gel and also protrudes from the tubes. Make sure you mark which side of the cell was soaked in which solution.

Part II: Testing the Cells at Room Temperature

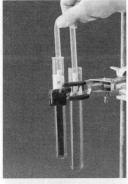

Place 30 mL of each of the salt solutions that correspond to the metals being used into large test tubes. Label the test tubes accordingly. Use a ring stand to hold the two test tubes next to each other.

Insert the appropriate salt bridge into the two tubes. Make sure that you place the correct sides of soaked cotton in the correct solutions. Also make sure that the cotton is fully submerged in the salt solution.

Obtain a strip or piece of the metals being tested and clean the surface with the sandpaper provided.

Set the multimeter to read as a voltmeter in direct current volts (V_{DC}).

Using the clips on the voltmeter, connect the voltmeter to the metal strips and the sides of the test tubes. Make sure that the metal pieces are in contact with the solution but do not submerge the clips in the solution.

Measure the potential of the cell. If the potential is negative, reverse the wire connections with the voltmeter. Report your findings to the rest of your class.

Part III: Measuring the Potential at Different Temperatures

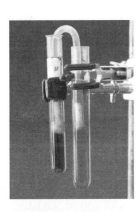

Set up your potential cell as you did in Part I.

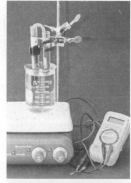

To the setup, add a 400 mL beaker half filled with DI water.
Place the beaker on a hot plate and adjust the cell's position on the ring stand until both test tubes are partially submerged in the water.

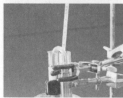

Set the hot plate temperature to 70 °C. Monitor the cell temperature until it reaches and maintains 70 °C. Measure the voltage.

Remove the cell from the hot plate and measure the voltage at 15 °C intervals until the cell cools to room temperature.

Create an ice-water bath in the same beaker as before and submerge the cell as before. Wait 10 min and record the temperature and the voltage.

Report Contents and Questions

The purpose should include a majority of the topics and techniques covered in the experiment and a description of the criteria for success. The procedure section should cite the lab manual and include any changes made to the procedure during lab work. Be sure to note your assigned metals.

The data section should include a table with the following: **(a)** experimental values for cell potentials for each of the 10 cells, **(b)** theoretical cell potentials based on literature values for each of the 10 cells, and **(c)** the percent error for the potentials of the 10 cells. A second table should be constructed to contain the data collected on your cell regarding its temperature dependence and should include the following: **(a)** the temperature at which each measurement was made, **(b)** the cell potential recorded at each temperature, **(c)** $\Delta G°$ at each temperature, **(d)** K at each temperature, and **(e)** ΔH and ΔS for the cell.

The calculation section should include a graph of ΔG versus temperature. All graphing guidelines apply. Example calculations of the following should also be included: percent error between the theoretical and experimental potentials for each cell, ΔG, and K. ΔG and K can be determined from the following equations:

$$\Delta G = -RT \ln K$$

and

$$E = (RT/\ln \mathfrak{I}) \ln K$$

The values for ΔH and ΔS are determined from the graph, where the slope of the graph is $-\Delta S$ and the y-intercept of the graph is ΔH. Make sure that the units are correct in all of your calculations and answers.

The conclusion should be several paragraphs with discussions of the values for ΔG, K, ΔH, and ΔS. Also, describe and discuss the potential series you developed from the class data and the discrepancies between the experimental cell potentials and the literature based cell potentials. Finally, discuss the relationship between temperature and cell potential. Remember to use values from your data and calculations to support your conclusions and discussions. As usual, report and explain all errors that may have occurred.

Answer the following questions:

1) What is a watt? How is it related to the potential of a battery?

Experiment 23
Laboratory Preparation

Name: _____ Date: _____

Instructor: _____ Sec. #: _____

Show all work for full credit.

1) Read the background, procedure, and report sections of the lab experiment carefully and develop a hypothesis of what information you expect to gain from the completion of the lab experiment.

2) Create any and all tables you might need to collect data during the experiment and then transfer those tables into your lab manual for use during the lab.

Experiment 23

Prelaboratory Assignment

Name: _____ Date: _____

Instructor: _____ Sec. #: _____

Show all work for full credit.

1) Using literature values for the half-cell reactions of the following metals, calculate the total cell potentials for each of the 10 possible cells. Assume complete reduction of the metals. **Copy the resulting potentials into your lab notebook.**

   Metals: Fe^{3+}, Cu^+, Mg^{2+}, Al^{3+}, Pb^{2+}

2) The standard potential of a galvanic cell ($E°_{cell}$) can be determined from thermochemical data such as the free energy of formation ($\Delta G°_f$) of each of the species in the reaction.

   For example, given the reaction:

   $$Fe^{3+}(aq) + Al(s) \leftrightarrow Al^{3+}(aq) + Fe(s)$$

   First, use the data to the right to calculate the $\Delta G°_{rxn}$. Then calculate $E°_{cell}$ from $\Delta G°_{rxn}$.

Species	$Fe^{3+}(aq)$	$Fe(s)$	$Al^{3+}(aq)$	$Al(s)$
$\Delta G°_f$ (kJ/mol)	−10.5	0	−481.2	0

3) Following is the shorthand notation for a Galvanic Cell:

$Al(s)|Al^{3+}(1\ M)||Cu^+(1\ M)|Cu(s)$

Write the overall balanced equation occurring in this cell, and indicate the physical states of reactants and products, that is, Al(s), Cu+(aq), and so on.

Identify the anode and cathode of this cell.

You make the following voltage measurements at four different temperatures.

Temperature (°C)	10	35	55	74
E°_{cell} (V)	2.190	2.176	2.165	2.154

Calculate the ΔG°_{rxn} at each of these temperatures.

Now, graph this data using the relationship $\Delta G = \Delta H - T\Delta S$ and determine ΔH°_{rxn} (in kJ/mol) and ΔS°_{rxn} (in J/[K · mol]) for this process. (This assumes that ΔH°_{rxn} and ΔS°_{rxn} do not vary with temperature. Remember to use absolute temperature and pay attention to units and signs in your plot and your calculations.)

Experiment 24
Qualitative Analysis: Anions

Introduction

Qualitative analysis refers to processes used to determine the presence of a chemical substance based on its reactivity. As the name indicates, qualitative analysis is not concerned with the amount of substance present; rather, it is used to confirm its presence. As modern scientific equipment has improved, wet chemistry qualitative analysis techniques have given way to instruments such as the gas chromatograph, infrared spectrometer, and mass spectrometer for analysis of unknown mixtures. These instruments have the advantage of indicating not only what is present but also the amount present. However, wet chemistry qualitative analysis is still important for two reasons: (1) speed and mobility and (2) developing a firm understanding of general chemical reactivity.

The first of the two reasons, speed and mobility, refers to the need for fast, on-site analysis. The instruments mentioned above are predominantly laboratory bound and thus inefficient if there is a need to analyze a chemical substance in the field. If you have ever watched a CSI (crime scene investigation) television show, you have probably seen a character spray a solution on a suspect's hands or clothes to confirm gunshot residue; this is qualitative analysis.

The second of the two reasons goes to the very heart of what it means to be a chemist. To call yourself a chemist, you need to be able to predict whether a reaction will take place and, more importantly, the products of a reaction. A majority of qualitative analysis techniques require the knowledge you have already gained regarding solubility, acid–base chemistry, and redox reactions to determine which compounds will form under what conditions.

Background

Qualitative analysis is the process of determining the identity rather than the amount of chemical compounds (quantitative analysis). The qualitative procedure usually uses known reactions of a given chemical species and interprets the results using deductive reasoning. Qualitative analysis applies many aspects of chemistry, such as acid–base equilibria, redox reactions, and solubility. To increase the accuracy of the analysis process, the procedure generally breaks collections of unknown ions into groups of ions that follow the same chemical patterns. Groups are based on the reactivity of the ions, not on their groups in the periodic table. For example, anions with similar reactivities are all placed in the same group and then identified by use of chemical confirmatory reactions.

In this experiment, you will be given several known anion solutions on which to perform confirmation experiments. It is very important to take copious notes regarding the appearance and possibly the smell of the reacting anions. After you have completed observations of the known anion solutions, you will be given an unknown containing up to two of the anions previously tested. The reactivity of the anions supplied as known that may be found in the unknown is described below:

A) CO_3^{2-}

Only the alkali metal and ammonium carbonates are water soluble. Heating (at Bunsen burner temperatures > 500 °C) decomposes all but the alkali and alkaline earth metal carbonates, giving the oxide and carbon dioxide:

$$CuCO_3(s) \rightarrow CuO(s) + CO_2(g)$$

Dilute hydrochloric acid (H^+ in the reaction below) gives vigorous effervescence with carbonates, evolving carbon dioxide:

$$CO_3^{2-}(aq \text{ or } s) + 2H^+(aq) \rightarrow H_2O(l) + CO_2(g)$$

B) SO_4^{2-}

$BaSO_4$, $SrSO_4$ and $PbSO_4$ are insoluble. $CaSO_4$ is only slightly soluble. Barium chloride solution added to sulfate solution acidified with dilute hydrochloric acid produces a white precipitate of barium sulfate:

$$BaCl_2(aq) + H_2SO_4 (aq) \rightarrow BaSO_4(s) + 2HCl(aq)$$

Adding lead(II) acetate solution gives a precipitate of white lead sulfate:

$$Pb(C_2H_3O_2)_2(aq) + H_2SO_4(aq) \rightarrow PbSO_4(s) + 2HC_2H_3O_2(aq)$$

C) PO_4^{3-}

A test for phosphate ion in solution is performed by adding nitric acid followed by a solution of ammonium molybdate, $(NH_4)_6Mo_7O_{24}$ **In acidic solution, the ammonium molybdate forms molybdic acid, which reacts with the phosphate ion.** If phosphate ion is present, a canary-yellow precipitate of triammonium dodecamolybdophosphate will form when heated:

$$Ca_3(PO_4)_2(s) + 6H^+(aq) \rightarrow 3Ca^{2+}(aq) + 2H_3PO_4(aq)$$

$$PO_4^{3-} + 12H_2MoO_4 + 3NH_4^+ \rightarrow (NH_4)_3[PO_4 \bullet 12MoO_3] \bullet 12H_2O$$

D) I⁻

AgI, PbI$_2$, Hg$_2$I$_2$, and CuI are all insoluble in water. When added to solid iodides, concentrated sulfuric acid gives a mixture of hydrogen iodide, iodine, hydrogen sulfide, sulfur, and sulfur dioxide; the HI produced is oxidized by sulfuric acid. The mixture evolves purple acidic fumes, turns into a brown slurry, and is a mess, but it does confirm the presence of iodide:

$$NaI(s) + H_2SO_4(l) \rightarrow NaHSO_4(s) + HI(g)$$

$$2HI + H_2SO_4 \rightarrow I_2 + SO_2 + 2H_2O$$

$$6HI + H_2SO_4 \rightarrow 3I_2 + S + 4H_2O$$

$$8HI + H_2SO_4 \rightarrow 4I_2 + H_2S + 4H_2O$$

NOTE: There are no state symbols in these equations because the mixture is such a mess and is more sulfuric acid than water.

Addition of silver nitrate solution to a solution of an iodide that has been acidified (test with blue litmus paper) with dilute nitric acid gives a yellow precipitate of silver iodide. The precipitate is insoluble even in concentrated ammonia:

$$AgNO_3(aq) + HI(aq) \rightarrow AgI(s) + HNO_3(aq)$$

$$AgI(s) + NH_3(aq) \rightarrow \text{No reaction}$$

Acidification with nitric acid is necessary to eliminate carbonate or sulfite, both of which interfere with the test by giving spurious precipitates. An alternative test for iodine, introduced in the halogens experiment, is to use an oxidizing agent to oxidize iodide to iodine, which is brown in aqueous solution. A suitable oxidizing agent is sodium hypochlorite; this is added to the test solution, followed by a little dilute hydrochloric acid and a few milliliters of hexane:

$$2NaOCl(aq) + 2HCl(aq) + 2HI(aq) \rightarrow I_2(aq) + 2NaCl(aq) + 2H_2O(l)$$

Iodine can be extracted from the solution by shaking with an immiscible organic solvent, such as hexane, and noting the purple color of the organic layer.

E) Cl⁻

The chloride salts AgCl, PbCl$_2$, Hg$_2$Cl$_2$, and CuCl are all insoluble in water. Concentrated H$_2$SO$_4$ produces steamy acidic fumes of HCl from solid chlorides:

$$NaCl(s) + H_2SO_4(aq) \rightarrow NaHSO_4(s) + HCl(g)$$

Addition of silver nitrate solution to a solution of a chloride that has been acidified (test with pH paper) with dilute nitric acid gives a white precipitate of silver chloride. The precipitate is readily soluble in dilute ammonia or in sodium thiosulfate solution:

$$AgNO_3(aq) + HCl(aq) \rightarrow AgCl(s) + HNO_3(aq)$$

$$AgCl(s) + 2NH_3(aq) \rightarrow [Ag(NH_3)_2]^+(aq) + Cl^-(aq)$$

$$AgCl(s) + 2S_2O_3{}^{2-}(aq) \rightarrow [Ag(S_2O_3)_2]^{3-}(aq) + Cl^-(aq)$$

Acidification with nitric acid is necessary to eliminate carbonate or sulfite, both of which interfere with the test by giving spurious precipitates.

Concentrated solutions of sulfates can give a precipitate of silver sulfate in this test; its appearance is wholly different from AgCl. The latter is a flat white; the sulfate is a pearly white, rather like pearlescent nail polish.

F) CH₃COO⁻

Acetates on heating with dilute hydrochloric acid give acetic acid, recognizable by its vinegary smell. Neutral iron(III) chloride solution added to neutral solutions of acetate ion gives a deep red color owing to formation of iron(III) acetate.

$$NaCH_3COO(aq) + HCl(aq) \rightarrow CH_3COOH(aq) + NaCl(aq)$$

$$3NaCH_3COO(aq) + FeCl_3(aq) \rightarrow Fe(CH_3COO)_3(aq) + 3NaCl(aq)$$

G) Br⁻

AgBr, PbBr$_2$, Hg$_2$Br$_2$, and CuBr are all insoluble in water. Concentrated sulfuric acid gives a mixture of hydrogen bromide, bromine, and sulfur dioxide with solid bromides; the HBr produced is oxidized by sulfuric acid. The mixture evolves steamy brownish acidic fumes:

$$NaBr(s) + H_2SO_4(l) \rightarrow NaHSO_4(s) + HBr(g)$$

$$2HBr(aq) + H_2SO_4(aq) \rightarrow Br_2(aq) + SO_2(g) + 2H_2O(l)$$

Addition of silver nitrate to a solution of a bromide that has been acidified (test with litmus paper) with dilute nitric acid gives a cream precipitate of silver bromide. The precipitate is readily soluble in concentrated ammonia:

$$AgNO_3(aq) + HBr(aq) \rightarrow AgBr(s) + HNO_3(aq)$$

$$AgBr(s) + 2NH_3(aq) \rightarrow [Ag(NH_3)_2]^+(aq) + Br^-(aq)$$

Acidification with nitric acid is necessary to eliminate carbonate or sulfite, both of which interfere with the test by giving spurious precipitates. Oxidizing agents oxidize bromide to bromine, which is yellow or orange in aqueous solution. Bromine can be extracted from the solution by shaking with an immiscible organic solvent, such as hexane, and noting the orange color of the organic layer.

A suitable oxidizing agent is sodium hypochlorite; this is added to the test solution, followed by a little dilute hydrochloric acid and a few milliliters of hexane:

$$2NaOCl(aq) + 2HCl(aq) + 2HBr(aq) \rightarrow Br_2(aq) + 2NaCl(aq) + 2H_2O(l)$$

H) NO_3^-

Since all nitrates are water soluble, there is no precipitation reaction for this ion. In chemical analysis, a test for nitrates involves the addition of a solution of iron(II) sulfate to the substance to be tested, followed by the addition (without mixing) of a few drops of concentrated sulfuric acid; the presence of a nitrate is indicated by the formation of a brown ring where the sulfuric acid contacts the test mixture. The brown ring is a complex containing the $Fe(NO)^{2+}$ ion.

I) SO_3^{2-}

A sulfite solution acidified with dilute HNO_3 produces sulfur dioxide gas on warming; bubbling of SO_2 gas through potassium dichromate(VI) solution produces a green colored solution.

$$Na_2SO_3(aq) + 2HNO_3(aq) \rightarrow H_2O(l) + SO_2(g) + 2NaNO_3(aq)$$

J and K) CrO_4^{2-} and $Cr_2O_7^{2-}$

These ions are related through the equilibrium

$$Na_2Cr_2O_7(aq) + 2NaOH(aq) \Leftrightarrow 2Na_2CrO_4^{2-}(aq) + H_2O(l)$$

In alkaline solution, the yellow chromate dominates and in acidic solution, orange dichromate dominates. All dichromate salts are soluble; addition of dichromate ions to solutions of ions of metals that have insoluble chromate salts leads to the precipitation of chromates.

Addition of barium chloride solution to a chromate or dichromate solution precipitates bright yellow barium chromate:

$$BaCl_2(aq) + Na_2CrO_4(aq) \rightarrow BaCrO_4(s) + 2NaCl(aq)$$

Acidified potassium dichromate solutions oxidize primary alcohols to aldehydes and then acids, and secondary alcohols to ketones. Ethanol can be used to test for dichromate; it turns the solution green, and the apple smell of ethanal is evident.

$$K_2Cr_2O_7(aq) + 8H^+(aq) + 5H_2O(l) + CH_3CH_2OH(aq) \rightarrow CH_3CH{=}O(l) + 2[Cr(H_2O)_6]^{3+}(aq)$$

The orange dichromate(VI) ion is reduced to hexaquachromium(III) in this reaction.

Procedure

REMINDERS: Solution volumes can be estimated using the guideline that 20 drops is ~1 mL.

Be sure you are using the correct concentrations of each reagent!

Part I

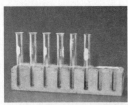

Collect a 1 mL sample of each known anion solution and place each in a labeled test tube. Conduct the appropriate test on each anion and take careful notes on the appearance, pH, and smell (when appropriate) of the anion solution before, during, and after reaction.

Test for *carbonate*

Cautiously add 1 mL of 1 M hydrochloric acid to 1 mL of your known salt solution. Odorless bubbles will be noted if carbonate is present.

Test for *sulfate*

Cautiously add 1 mL of 6 M hydrochloric acid to 1 mL of your known salt solution. Next, add a few drops of 1 M barium chloride. A white, finely divided precipitate indicates the presence of the sulfate ion.

Test for *phosphate*

Cautiously add 1 mL of 6 M nitric acid to 1 mL of your known salt solution. Next, add 2 or 3 mL of 0.5 M ammonium molybdate and stir. A yellow precipitate indicates the presence of phosphate. Set the solution aside to observe later, as this precipitate may be slow to form.

Test for *iodide*

Dip a piece of starch packing material into a small amount of your known salt solution. Place the packing material into a test tube and add 1 mL of dilute hydrogen peroxide. A yellow bubbly substance indicates the presence of iodide.

Test for *chloride*

Cautiously add 1 mL of 6 M nitric acid to 1 mL of your known salt solution. Next, add a few drops of 0.1 M silver nitrate. A white precipitate indicates the presence of chloride.

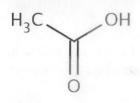

Test for *acetate*

Cautiously add 1 mL of 3 M sulfuric acid to 1 mL of your known salt solution and stir. A vinegar odor is indicative of acetate.

Test for *bromide*

Cautiously add drops of 6 M nitric acid to 1 mL of your known salt solution until the solution is acidic, testing with pH paper. Add 10 drops of hexane and 5 drops of 0.1 M $KMnO_4$ to the solution. Shake the test tube by tapping gently. The hexane layer turning rust brown indicates the presence of bromide.

Test for *nitrate*

Acidify 1 mL of nitrate solution by adding 3 M H_2SO_4 as needed. Next, add 10 drops of a freshly prepared, saturated solution of iron(II) sulfate ($FeSO_4$) and mix gently. Incline the test tube at a 45° angle and carefully add 10 drops of concentrated H_2SO_4 so the drops roll down the side of the test tube and slide gently onto the top of the solution. DO NOT MIX the solutions! Two separate liquid layers will be observed in the test tube. If NO_3^- is present, a *very faint* brown ring will be observed near the bottom of the test tube, thereby confirming the presence of nitrate ion. Set the tube aside to observe again later, as the ring may be slow to form.

Brown Ring

Test for *sulfite*

Add 10 drops of aqueous barium nitrate to 1 mL of your known salt solution. A white precipitate should form. Add 5 drops of 0.1 M potassium dichromate. Shake by tapping the test tube. Now, add 3 M HNO_3 dropwise until the precipitate is dissolved. A change of the orange color of the solution to green confirms the presence of sulfite.

301

Test for *chromate*

Add 10 drops of aqueous barium chloride to 1 mL of your known salt solution. A yellow precipitate should form, confirming the presence of chromate.

Test for *dichromate*

Add 1 mL of 95% ethanol to 1 mL of your known salt solution. Change of the solution color to green and the production of a "fruity" odor confirm the presence of dichromate.

Part II

Obtain an unknown solution of anions from your instructor. Note the identity code of the unknown in your notebook. Using the procedures and results from the tests done previously on the known anions, determine the two anions in your unknown. Make sure you can support your conclusions by showing negative test results for the anions that you believe are not in your unknown.

Report Contents and Questions

The purpose should include a majority of the topics and techniques covered in the experiment and a description of the criteria you will use to determine its success or failure. The procedure section should cite the lab manual and include any changes made to the procedure during lab work. The procedure section should also include a step-by-step description of how you determined the identity of the unknown.

The data section should include a table for Parts I and II. In the first column, record initial observations, noting the color of each solution. In the remaining sections, describe the results upon mixing. Be sure to include things such as colors and formulas of precipitates. DO NOT RECORD "NO CHANGE"; RECORD WHAT THE SOLUTION/MIXTURE LOOKS LIKE EVEN IF IT IS CLEAR AND COLORLESS. For the anion precipitation column, indicate which anions form a precipitate and all other observations. **All observations written in this table must match those made during lab and recorded in your lab notebook.**

For Part II, include observations made during each step you took in lab while testing your unknown, along with the code and identity of the anion(s) in your unknown.

For the calculation section for this experiment, write the balanced net ionic equation for each reaction from Part I.

The conclusion section should include several detailed paragraphs, with both the identities of the ions in your unknown and how you reached these conclusions. This should include a discussion of the known reactions you did in Part I and how they compare to the unknown reactions. Be sure to include a description of both supporting and eliminating experiments that you conducted. Finally, include all possible errors found in this experiment.

Answer the following questions:

1) If you were given an unknown solution that you were told contained either Cl^- or I^-, both anions, or neither, explain how you would confirm or deny the presence or absence of the ions.

2) Why are the reagents used to test for anions usually a nitrate of the cation that is reacting rather than other salts of that cation?

Experiment 24
Laboratory Preparation

Name: _____ Date: _____

Instructor: _____ Sec. #: _____

Show all work for full credit.

1) Read the background, procedure, and report sections of the lab experiment carefully and develop a hypothesis of what information you expect to gain from the completion of the lab experiment.

2) Create any and all tables you might need to collect data during the experiment and then transfer those tables into your lab manual for use during the lab.

Experiment 24
Prelaboratory Assignment

Name: _____ **Date:** _____

Instructor: _____ **Sec. #:** _____

Show all work for full credit.

1) Tell how you could distinguish chemically between the following pairs of substances in one or two steps. In each case, tell what you would expect to observe and include equations for reactions occurring.
   a. HCl and H_2SO_4
   b. NaBr and Na_2CO_3
   c. $BaSO_4$ and $BaCO_3$
   d. KNO_3 and K_3PO_4

2) You are given three unlabeled containers and are told that they contain nitric acid, sulfuric acid, and phosphoric acid. How could you determine which acid is in which container?

Experiment 25
Qualitative Analysis: Cations

Introduction

Qualitative analysis refers to the methods used to determine the identity of a chemical species as opposed to its amount. Most of your lab work this semester has been spent determining how *much* of a particular analyte is present through a variety of quantitative techniques. In this experiment, instead of concentrating upon how *much* of something is present, you will be concerned only with *what* is present. The qualitative procedure uses known reactions of a given chemical species and interprets the results using deductive reasoning. Your understanding of acid–base equilibria, redox reactions, and solubility will be enhanced by looking at why each of the reactions works.

To increase the accuracy of the analysis, the procedure for a typical qualitative assay generally separates ions into groups that follow similar chemical patterns. It is important to remember that the groups are formed on the basis of the reactivity of the ions, not their groups in the periodic table! Specific ions are then identified by use of confirmatory reactions.

Background

Qualitative analysis is used to separate and identify the cations and anions present in an unknown chemical mixture. Ions are grouped according to their reactivity with specific compounds. First, ions are separated into groups from the initial aqueous solution using their differences in solubility. After each group has been separated, testing is conducted for the individual ions in each group. What follows is a common grouping of cations.

Group I cations are those that form insoluble chlorides when reacted with dilute HCl solutions. This group includes silver (Ag^+), mercury(I) (Hg_2^{2+}), and lead(II) (Pb^{2+}), which are precipitated out by addition of HCl to the unknown cation mixture. The Group I cations form the precipitates $AgCl$, Hg_2Cl_2, and $PbCl_2$ that can then be removed from the mixture by centrifugation or vacuum filtration.

Group II cations, while soluble in dilute HCl, are insoluble when reacted with a dilute H_2S solution at low (acidified) pH. Group II cations include bismuth(III) (Bi^{3+}), cadmium(II) (Cd^{2+}), copper(II) (Cu^{2+}), antimony(III) (Sb^{3+}), antimony(V) (Sb^{5+}), tin(II) (Sn^{2+}), and tin(IV) (Sn^{4+}). The Group II cations are precipitated out by reaction with H_2S at a low pH, normally around 0.5. The precipitates that are formed are removed from the remaining mixture by centrifugation.

Group III cations are soluble in acidic H_2S but are insoluble in basic H_2S. Group III cations, which include aluminum (Al^{3+}), chromium(III) (Cr^{3+}), iron(II) (Fe^{2+}), iron(III) (Fe^{3+}), manganese(II) (Mn^{2+}), and zinc(II) (Zn^{2+}), are precipitated out of the mixture by changing the pH of the previous solution (already saturated with H_2S) to a pH of 9 or greater.

All that is left in the unknown solution at this point should be the Group IV and V cations. These are the alkali and alkaline earth metal cations and the ammonium cation. These cations can be separated from each other by their varying reactions with carbonate solution.

Groups IV and V cations include barium (Ba^{2+}), calcium (Ca^{2+}), magnesium (Mg^{2+}), and strontium (Sr^{2+}), and potassium (K^+), sodium (Na^+), and ammonium (NH^{4+}), respectively. The alkaline earth metal cations Ba^{2+}, Ca^{2+}, Mg^{2+}, and Sr^{2+} are precipitated in $(NH_4)_2CO_3$ solution at a pH of 10 and then removed from the others by centrifugation. The alkali metal cations K^+ and Na^+ are soluble in almost all solutions and thus cannot be separated by precipitation. Rather, their presence is confirmed by a flame test. The ammonium ion is also soluble and may also have been lost as ammonia through earlier tests. To test for the presence of ammonium ion, NaOH is added to a small amount of the original mixture until the mixture becomes basic. At that point, the smell of ammonia (NH_3) will indicate the presence of ammonium ion (NH^{4+}), and confirmation can be made by use of moist red litmus paper held at the mouth of the test tube.

The portion of qualitative analysis described above simply separates the groups from each other. To determine the presence of individual ions in each group, a variety of other reactions must be performed. The selection of ions that will be used in this experiment include Group I: Ag^+ and Pb^{2+}; Group II: Cu^{2+} and Sn^{4+}; Group III: Mn^{2+} and Al^{3+}; Group IV: Ba^{2+} and Sr^{2+}; and Group V: Na^+, K^+, and NH_4^+. Each reaction is specifically designed to react with only one cation in each group, thereby confirming only its presence. As with the main group separations, the reactions with each cation rely on those chemical properties that are unique to that ion. The discussion of the separation and confirmation of those ions is quite extensive and best broken up into several parts. The paragraphs below provide more detailed information about the reactions used to confirm the presence of each cation above listed by group.

Group I Cation Confirmation Tests

The result of the separation of the Group I cations from the other groups is a precipitate containing all of the Group I cations. To analyze the Group I cations further, the precipitate needs to be separated. To do this, you must know something about the reactivities of the two Group I cations, silver (Ag^+) and lead(II) (Pb^{2+}).

Lead(II) chloride ($PbCl_2$) is much more soluble than silver chloride (AgCl) and thus can be redissolved by addition of hot distilled water (remember, solubility is temperature dependent, and most substances are more soluble at higher temperatures). Lead(II) chloride is almost three times as soluble in hot water as in cold water. Centrifugation of the resulting solution allows you to separate the lead(II) cations from the silver cations. At this point, you should have a supernatant containing lead(II) and a precipitate containing silver cations.

For confirmation of the presence of lead(II), we need a reaction that generates a colored precipitate that is characteristic of lead. Addition of several drops of potassium chromate (K_2CrO_4) to the solution containing lead(II) will produce a bright yellow precipitate, confirming the presence of the lead. The total ionic reaction is:

$$PbCl_2(aq) + K_2CrO_4(aq) \rightarrow 2KCl(aq) + PbCrO_4(s) \downarrow \text{ppt}$$

The net ionic reaction is:

$$Pb^{2+}(aq) + CrO_4^{2-}(aq) \rightarrow PbCrO_4(s) \downarrow \text{ppt}$$

The remaining precipitate is then treated with aqueous ammonia. The silver cations in the precipitate react with ammonia to form a complex called the diamminesilver(I) complex:

$$AgCl(s) + 2NH_3(aq) \rightarrow Ag[(NH_3)_2]^+(aq) + Cl^-(aq)$$

This complex is the principal ingredient in Tollen's reagent, known as the silver-mirror test. It is colorless and is used to distinguish between an aldehyde or ketone. An aldehyde is an organic compound containing carbon, hydrogen, and oxygen formed by the oxidation of alcohols. A ketone is also an organic compound; however, it contains a carbonyl group (carbon double bonded to an oxygen) bonded to two hydrocarbons. Addition of strong nitric acid solution (HNO_3) to the resulting solution will break down the diamminesilver(I) complex. Once the complex is broken, the silver can again react with the chloride ions still present in the solution and precipitate as AgCl. The presence of a white precipitate thus confirms the presence of the Ag^+ cation:

$$Ag[(NH_3)_2]^+(aq) + 2H_3O^+(aq) + Cl^-(aq) \rightarrow AgCl(s) \downarrow + 2NH_4^+(aq) + 2H_2O(l)$$

Group II Cations

The supernatant, the top portion of the solution, not containing the solids/precipitates, that remained after the removal of the Group I cations contains the Group II through V cations. To separate and confirm the Group II cations, copper(II) (Cu^{2+}) and tin(IV) (Sn^{4+}), we need to know only that they, unlike the rest of the remaining ions, are insoluble in acidified sulfide solution.

NOTE: Hydrogen sulfide (H_2S) is used to precipitate the Group II cations. However, H_2S is a dangerous gas in its pure form. It is much safer to use thioacetamide (CH_3CSNH_2) to form H_2S in solution. When heated, aqueous solutions of thioacetamide hydrolyze to produce H_2S. The formation of H_2S from thioacetamide also provides another level of control in separation of the group II ions. Below are three reactions based on different solution conditions.

In acidic solution:

$$CH_3CSNH_2 + 2H_2O + H^+ \rightarrow CH_3COOH + NH_4^+ + H_2S$$

In basic solution using ammonia:

$$CH_3CSNH_2 + 2H_2O + 2NH_3 \rightarrow CH_3COO^- + 3NH_4^+ + S^{2-}$$

In basic solution using strong base:

$$CH_3CSNH_2 + 2OH^- \rightarrow CH_3COO^- + NH_3 + S^{2-} + H_2O$$

Each of these reactions produces different concentrations of sulfide ion. An acidic solution produces a lower concentration of sulfide, which will precipitate all of the very insoluble Group II sulfides, leaving any slightly insoluble sulfides in solution. A strong base solution produces the most sulfide and is capable of precipitating the least insoluble sulfides. Thus, by controlling pH, you can determine which cations precipitate.

To precipitate the Group II cations out of the supernatant, we must first acidify the solution to a pH of ~0.5. This is done by adding HCl to the supernatant and confirming the pH by use of pH paper. The thioacetamide (CH_3CSNH_2) reacts with the acid and water to form H_2S, which in turn precipitates the group II cations:

$$Cu^{2+}(aq) + S^{2-}(aq) \rightarrow CuS(s) \downarrow$$
$$Sn^{2+}(aq) + S^{2-}(aq) \rightarrow SnS(s) \downarrow$$

The precipitate is removed from solution by centrifugation, and the supernatant containing the remainder of the Groups III through V ions is saved. The precipitate should be washed with HCl at least once to remove any remaining group II cations and the supernatant from those washes combined with the first. Once you are satisfied that all of the group II cations have been removed, you can begin to separate the cations from each other.

Separation and confirmation of tin(IV): Unlike copper, tin is soluble in basic solution. Addition of KOH until the solution is basic should dissolve any tin present as the $Sn(OH)_6^{2-}$ complex ion. The solution is heated, centrifuged while still hot and the supernatant containing the tin complex is removed. Tin(IV) is then confirmed by the addition of HCl until the solution becomes acidic followed by the addition of more thioacetamide. The solution is heated for 5+ minutes and the development of a yellow precipitate (SnS_2) confirms the presence of tin.

$$Sn(OH)_6^{2-}(aq) + 6H^+(aq) + 6Cl^-(aq) \rightarrow SnCl_6^{2-}(aq) + 6H_2O(l)$$
$$Sn^{4+}(aq) + 2S^{2-}(aq) \rightarrow SnS_2(s)$$

The precipitate that was saved from above now contains only the copper(II) (Cu^{2+}) cations.

Confirmation of copper(II): The precipitate is redissolved using nitric acid and heat. If copper is present, the solution should be a pale blue color due to the presence of $Cu(H_2O)_4^{2+}$. To confirm the presence of copper(II), add acetic acid and then some potassium ferrocyanide ($K_4[Fe(CN)_6]\cdot3H_2O$). The formation of a reddish brown precipitate confirms the presence of copper(II).

Group III Cations

Group III consists of manganese(II) (Mn^{2+}) and aluminum (Al^{3+}). Oxidizing the Group III cations with basic H_2O_2 ensures that $Mn(OH)_2$ and $Al(OH)_3$ precipitates are produced. Once the Group III cations are separated by centrifugation, treatment of the precipitate with excess base allows the amphoteric Al^{+3}, meaning they are able to react with both an acid and a base, to form a soluble complex with OH^-. This solubility difference permits the separation of the Mn ions from the Al ions:

$$Mn^{2+}(aq) + 2OH^-(aq) \rightarrow Mn(OH)_2(s) \text{ white solid}$$
$$3Al^{3+}(aq) + 4OH^-(aq) \rightarrow Al(OH)_4^-(aq) \text{ clear solution}$$

Analysis of Mn^{2+}: $Mn(OH)_2$ is solubilized in HNO_3 and H_2O_2 to dissolve the salt, and the resulting aqueous sample is tested for the presence of aqueous Mn^{2+} by addition of $NaBiO_3$.

$$Mn^{2+}(aq) + H_2O(l) + BiO_3^-(aq) \rightarrow MnO_4^-(aq) + 2H^+(aq) + Bi(s) \text{ purple solution}$$

The MnO_4^- ion generated in the reaction solution is purple and easily detected. The generation of the color is the confirmation of the presence of the Mn^{2+} ion in the sample.

Analysis of the Al^{3+}: The supernatant separated previously contains $Al(OH)_4^-(aq)$. Acidification with HNO_3 and heating eliminates the basic complexes and generates free Al^{3+} ions, which are amphoteric.

$$Al(OH)_4^-(aq) + 4H^+(aq) \rightarrow Al^{3+}(aq) + 4H_2O(l)$$

Confirmation of Al^{3+}: The aqueous Al^{3+} ion is then precipitated again as the $Al(OH)_3$ – aluminon dye complex.

$$Al(OH)_3(s) + 3CH_3CO_2H(aq) \rightarrow Al^{3+}(aq) + 3H_2O(l) + 3CH_3CO_2^-(aq)$$
$$Al^{3+}(aq) + 3OH^-(aq) + \text{aluminon dye }(aq) \rightarrow Al(OH)_3 - \text{aluminon dye}(s) \text{ red solution/solid}$$

This step is the confirmation of the presence of Al^{3+}. The dye is added with NH_4Cl solution, which is buffered with NH_3. A pH of -5.7 or so ensures that only the correct equivalent amount of OH^- is added. Too much OH^- will cause $Al(OH)_4^{-1}(aq)$ to form and also yield a negative test with the dye. The dye itself is needed to give the $Al(OH)_3(s)$ a color, since its presence might be missed otherwise. Careful control of the pH is therefore necessary to ensure that a false negative result does not occur.

Group IV Cations

Once Groups I through III have been removed, all that should remain in an unknown sample are the Group IV and Group V cations. The Group IV cations are the alkaline earth metals Ba^{2+} and Sr^{2+}. To separate the Group IV cations from those of Group V, we note that all of the alkaline earth metals form insoluble carbonates when treated with ammonium carbonate $(NH_4)_2CO_3$. The precipitate that forms is removed by centrifugation and the supernatant saved, as it contains the Group V cations.

$$Ba^{2+}(aq) + CO_3^{2-}(aq) \rightarrow BaCO_3(s)$$
$$Sr^{2+}(aq) + CO_3^{2-}(aq) \rightarrow SrCO_3(s)$$

The precipitate containing the Group IV cations is redissolved by addition of acetic acid and heating.

Test for barium: Addition of potassium chromate (K_2CrO_4) will precipitate any barium present as a bright yellow solid. Centrifugation removes the precipitate, and the supernatant is saved to test for strontium.

$$Ba^{2+}(aq) + CrO_4^{2-}(aq) \rightarrow BaCrO_4(s) \quad \text{bright yellow precipitate}$$

Test for strontium: The supernatant is used to perform a flame test for strontium. A crimson-red flame confirms the presence of strontium.

Group V Cations

The supernatant removed from the separation of the Group IV cations contains the Group V cations (Na^+, K^+, and NH_4^+). These cations are soluble in most common reagents and thus cannot be confirmed by precipitation. Flame tests are therefore used to confirm both sodium and potassium. The sodium flame is a strong yellow color, and the potassium emits a weaker violet color. The potassium flame is observed through a piece of cobalt glass to remove the stronger sodium wavelengths that might otherwise mask its presence.

A sample of the original unknown mixture, not the supernatant above, is used to test for ammonium ion because several of the separation techniques used previously may have removed the ammonium as NH_3. A small amount of the original sample is made basic by use of 1 M of NaOH solution and heated. The odor of ammonia indicates the presence of the ammonium ion.

$$NH_4^+(aq) + OH^-(aq) \xrightarrow{\Delta} NH_3(g) + H_2O(l)$$

A piece of moistened red litmus paper held over the mouth of the test tube confirms the presence of the base.

Procedure

SAFETY: All of your tests should be performed in the fume hood, since many of the reactions produce gases. Be sure to dispose of all waste solutions properly according to your instructor's directions.

NOTE: You should keep the test tubes of all confirmation tests until you complete the experiment. This way, you can reinterpret your results if necessary.

First, start a hot water bath. It should contain distilled water. The bath should be boiling gently when used.

Unknown Mixture of Cations

Obtain your unknown mixture, which contains a subset of the 11 possible cations. Make sure to write down the code for the unknown as well as a description of the mixture such as color and pH (from indicator paper). Your task is to identify the cations in the mixture. Use your observations and the flowchart you prepared in your lab preparation to confirm the presence or absence of each cation. Check your flowchart with your instructor before beginning your analysis.

To confirm your analysis of the unknown mixture, use the known solutions of single cations to observe positive confirmation tests as necessary.

Part I: Separation and Analysis of Group I Ions

Group I ions are separated from the other cations in the unknown by addition of HCl, precipitation of the Group I metal chlorides, and separation of the solid precipitate from the remainder of the solution, which contains the dissolved Groups II through V ions. Be careful to avoid a large excess of HCl, since both cations form soluble complex cations with excess chloride ions.

1) Add 12 drops of 6 M HCl to no more than 2 mL of the solution to be analyzed. If a precipitate forms, centrifuge the sample and remove and save the liquid solution (supernatant). Make sure the solution is completely clear.

2) Add an additional drop or two of HCl to the supernatant and let the solution sit to see if more precipitate forms. If so, remove by centrifugation. Discard this additional precipitate (*be careful not to discard the precipitate from the earlier centrifugation*) and combine the supernatant with the supernatant from the first step for further analysis in Part II.

3) Carefully wash the precipitate with DI water and test it for lead and silver.

To make the most efficient use of time, you should begin the separation of Group II ions, for Part II, at this point. While that solution is heating in the water bath, you can return to your Group I ions and perform the confirmation tests for lead, mercury, and silver. Be sure to mark your test tubes carefully so that you will know which test tubes are yours and what they contain.

A. Confirmation of Lead (Pb^{2+})

Add ~40 drops of hot distilled water to the solid precipitate from above. Centrifuge while hot, decant, and save both the solution and the solid. Add 4 to 6 drops of 1 M K_2CrO_4 to the solution; a bright yellow precipitate confirms the presence of Pb^{2+}.

B. Confirmation of Silver (Ag^+)

Add ~10 drops of 6 M $NH_3(aq)$ to the solid from A. The solid should dissolve, but if any precipitate remains, centrifuge and proceed using only the supernatant. Acidify the solution by adding 6 M HCl, using pH indicator paper to test for a pH of ~0.5. A white precipitate (AgCl) confirms the presence of Ag^+. (The Cl^- needed for precipitation will be present from the prior dissolution of AgCl and the additional HCl.)

Part II: Separation and Analysis of Group II Ions

The Group II ions are separated from those in Groups III through V by adding hydrogen sulfide to the mixture in acidic solution. The Group II ions precipitate as sulfides on reaction with S^{2-} from H_2S.

1) Add 4 drops of 6 M HCl to the solution from Section 1 and dilute to ~5.0 mL using DI water. Confirm acidity with pH paper.

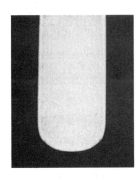

2) Add 20 drops of 5% thioacetamide solution, stir, and heat in a water bath for at least 10 min.

3) Centrifuge, decant (pour off the supernatant), and save the supernatant for Groups III through V identification.

4) Wash the precipitate by stirring it with ~20 drops of 1 M HCl, centrifuge, and combine the supernatant with that from the previous centrifugation. It is essential to remove all solid Group II ions so that they will not interfere with the Groups III through V analysis later on.

A. Confirmation of Tin (Sn^{4+})

Add drops of 1 M KOH until the solution becomes distinctly basic according to pH paper tests. Heat the test tube for several minutes in the hot water bath. Centrifuge and decant while the solution remains hot, saving both the precipitate and the supernatant. Take the supernatant and add 1 M HCl until the solution becomes acidic (~pH 1) and then add 10 drops of 0.1 M thioacetamide. The formation of a yellow precipitate confirms the presence of tin(IV).

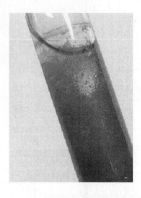

B. Confirmation of Copper (Cu²⁺)
To the precipitate from Part A, add 1 mL of 3 M nitric acid and heat in the hot water bath until the precipitate dissolves. Add 5 to 6 drops of potassium ferrocyanide. The production of a red-brown precipitate confirms the presence of copper(II).

Part III: Separation and Analysis of Group III Ions

The Group III ions form sulfides that are more soluble than those of Group II, but they may be precipitated as sulfides if the sulfide concentration is sufficiently large. Addition of concentrated NH₃(*aq*) to the thioacetamide ensures that the necessary sulfide concentration will be reached.

Your unknown solution should contain only Groups III through V ions after the procedure outlined in Part II above.

1) Add ~10 drops of 5% thioacetamide solution, stir well, and heat for ~5 min.

2) Add ~10 drops of concentrated aqueous NH₃, stir, and heat for 5 additional minutes. The precipitate contains the sulfides of the Group III ions. Centrifuge and decant the liquid. Save the supernatant for Groups IV and V identification.

A. Separation of Mn²⁺ from Al³⁺
Add another ~10 drops of concentrated aqueous NH₃ to the precipitate containing the Group III cations. Centrifuge and keep the supernatant for confirmation of the presence of aluminum.

B. Confirmation of Mn²⁺

Add 10 drops of 6 M HNO₃ and ~4 drops of 1 M sodium nitrite (NaNO₂) to the precipitate from Part A, and dilute to 2 mL. Heat the solution to fully dissolve the precipitate. Allow to cool and then add NaBiO₃ until the solution turns pink and then add ~4 to 6 drops or more of 6 M HNO₃. The formation of a transient pink to purple color of MnO_4^- confirms Mn^{2+}. An additional spatula tip of solid NaBiO₃ and a gentle shake of the test tube may make this subtle color change more apparent. Bismuthate (BiO_3^-) in strong acid is predominantly molecular bismuthic acid ($HBiO_3$) and is reduced to Bi^{3+}.

C. Confirmation of Al³⁺

To the solution set aside in Part A, add 5 drops of water and 2 drops of aluminon reagent and stir thoroughly. At this point, the solution may be colored because of the aluminon. Add 6 M aqueous NH₃ drop by drop, stirring well, until the solution is basic when tested with litmus paper. If Al^{3+} is present, the precipitate of $Al(OH)_3$ forms and adsorbs the aluminon from the solution, producing a red precipitate of $Al(OH)_3$. The supernatant is essentially colorless. The test for aluminum is not the red solution but rather the red precipitate, $Al(OH)_3$ and adsorbed aluminon dye. Centrifuge to concentrate the precipitate and check to see if the color of the precipitate is red.

Part IV: Separation and Analysis of Group IV Ions

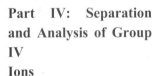

1) To the supernatant from Part III, add 30 drops of 3 M ammonium carbonate. Stir well.

2) Centrifuge for 2 min and decant. Save the precipitate. Set the supernatant aside for use in Group V analysis.

3) Wash the precipitate with DI water.

4) Centrifuge again and save the precipitate.

5) Add 16 drops of 6 M acetic acid and heat until all the precipitate dissolves.

6) Add 20 drops of 0.1 M K₂CrO₄ to the solution and stir well.

7) Heat the solution in a boiling water bath for 3 min. Centrifuge for

317

2 min and decant. Save the supernatant for confirmation of strontium.

A. Confirmation of Ba^{2+}

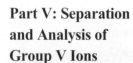

Wash the precipitate with DI water. Centrifuge and save the precipitate. Add 20 drops of 6.0 M HCl. Stir well and heat the solution if necessary until the precipitate dissolves. Perform a flame test by placing a small amount of the solution on a watch glass and vaporizing it over a Bunsen burner in the hood. A pale green flame confirms barium.

B. Confirmation of Sr^{2+}

Take the supernatant that was saved and perform a flame test as outlined above. A crimson-red flame confirms the presence of strontium.

Part V: Separation and Analysis of Group V Ions

All that should now remain in the supernatant saved from Part IV is the Group V ions, Na^+, K^+, and NH_4^+.

A. Confirmation of Na^+

Take the supernatant from Part IV and perform a flame test. The sodium flame is a strong yellow color.

B. Confirmation of K^+

The potassium emits a weaker violet color. The potassium flame is observed through a piece of cobalt glass to remove the stronger sodium wavelengths that might otherwise mask its presence.

C. Confirmation of NH_4^+

Obtain another small sample of the original unknown mixture in a test tube. Add ~20 drops of 1.0 M NaOH and heat in the hot water bath for a few minutes. The odor of ammonia confirms the presence NH_4^+ in the mixture. Dampened red litmus paper held at the mouth of the test tube during heating will change color in the presence of ammonia.

Report Contents and Questions

The purpose should include a majority of the topics and techniques covered in the experiment and the criteria for determining the success of the experiment. The procedure section should cite the lab manual and include any changes made to the written procedure during lab work. The procedure section should also include a step-by-step procedure on how you determined the identity of the unknown.

The data section should include an amended flowchart for the actual procedure and results you followed during the experiment. A table containing all of the observations you made to support your conclusions should be included. All observations written in this table must match those made during lab and recorded in your lab notebook.

The calculation section for this experiment will include balanced net ionic equations for each confirmation test you made in the determination of your unknown.

The conclusion section should include several detailed paragraphs with the identities of the ions in your unknown and how you reached these conclusions. This should also include a discussion of the known reactions you did for Groups I through V and how they compare to the unknown reactions. Be sure to include a description of both supporting and eliminating experiments that you conducted. Finally, include all possible errors found in this experiment.

Answer the following questions:

1) The confirmation test used in this experiment for copper produces a reddish-brown precipitate. There is another test that produces a blue precipitate. What is the difference in the tests?

2) Based on your knowledge of acid–base equilibria and solubility, describe why the solubility of different cation groups can be affected by the acidification of their solutions.

Experiment 25
Laboratory Preparation

Name: _____ Date: _____

Instructor: _____ Sec. #: _____

Show all work for full credit.

1) Read the background, procedure, and report sections of the lab experiment carefully and develop a hypothesis of what information you expect to gain from the completion of the lab experiment.

2) Create any and all tables you might need to collect data during the experiment and then transfer those tables into your lab manual for use during the lab.

Experiment 25

Prelaboratory Assignment

Name: _____ Date: _____

Instructor: _____ Sec. #: _____

Use the following flowchart to answer the questions below:

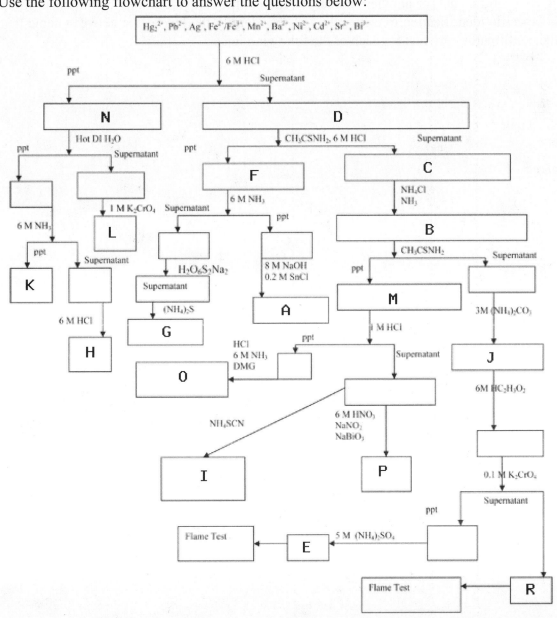

Experiment 25

1) Which box in the above scheme contains all of the ions, and only those ions, in each of the following groups?

Group I = _____

Group II = _____

Group III = _____

Group IV = _____

Group V = _____

2) Consult your lab manual, or another source, for the confirmatory reactions within each group that lead to specific identification of each cation. Then identify the box in the above scheme that contains the confirmatory product for each of the following ions.

Ag^+ = _____

Pb^{2+} = _____

Cd^{2+} = _____

Ba^{2+} = _____

Experiment 26
Qualitative Analysis: Cations, Anions, and Complex Ions

Introduction

Qualitative analysis is a process or processes used to determine the presence of a chemical substance based on its reactivity. As the name indicates, qualitative analysis is not concerned with the quantity of substance present; rather, it is simply used to confirm its existence. The overall process of a quals cheme has become somewhat historical in nature. As modern scientific equipment has improved, qualitative analysis techniques have given way to instruments such as the gas chromatograph and the infrared spectrometer or mass spectrometer for analysis of unknown mixtures. These instruments have the advantage of indicating not only what compounds are present in a sample (qualitative analysis) but also determining the concentration of those compounds (quantitative analysis). That being said, chemical qualitative analysis is still important for two reasons: (1) speed and mobility (2) every chemist needs a firm understanding of general chemical reactivity.

The first of the two reasons, speed and mobility, refers to the need for fast or on-site analysis. The equipment discussed above are generally laboratory bound and thus inefficient if there is a need to analyze a chemical substance in the field. If you have ever watched CSI (crime scene investigation) television shows, you have probably seen a character spray a solution on a suspect's hands or clothes to confirm gunshot residue. This is qualitative analysis.

The second of the two reasons goes to the very heart of what it means to be a chemist. To call yourself a chemist, you need to be able to predict whether a reaction will take place and more importantly, the products of that reaction. A majority of qualitative analysis techniques use the knowledge you have already gained regarding solubility, acid–base chemistry, and redox reactions to determine the compounds that will form and under what conditions.

Background

Introduction

Qualitative analysis refers to the methods used in determining the identity of a chemical compound as opposed to its amount (quantitative analysis). For example, a majority of this term has been spent determining how *much* of a particular analyte is present through a variety of quantitative techniques. However, in this experiment, we switch things up a little. Instead of concentrating on how *much* of something is present, you will only be concerned with *what* is present. Overall, the qualitative procedure uses known reactions of a given chemical species and interprets the results using deductive reasoning. In other words, a variety of chemical analyses exist for different elements or types of a compound, and when you combine these analyses with your understanding of acid–base equilibria, redox reactions, and solubility, it becomes fairly elementary to determine *what* particular analyte is found in an unknown mixture.

To increase the accuracy of the analysis process, the procedure for a typical qualitative assay generally breaks down several collections of unknown ions into groups of ions that follow the same chemical pattern. In fact, the **groups** are formed from the basis of the reactivity of the ions, not their groups in the periodic table! Similar ions are placed in the same group and are then identified by use of **confirmatory** reactions.

Identifying Anions and Cations

Qualitative analysis is used to separate and identify the cations and anions present in an unknown chemical mixture. Ions are grouped according to their reactivity with specific reagents. First, ions are removed in groups from the initial aqueous solution. After each group has been separated, testing is conducted for the individual ions in each group. What follows is a common grouping of cations.

Group I cations include silver (Ag^+), mercury I (Hg_2^{2+}), and lead II (Pb^{2+}) and are precipitated out by addition of 1 M HCl to the unknown cation mixture. The **Group I** cations form the precipitates $AgCl$, Hg_2Cl_2, and $PbCl_2$ that can then be removed from the mixture by centrifugation.

Group II cations include bismuth III (Bi^{3+}), cadmium II (Cd^{2+}), copper II (Cu^{2+}), antimony III (Sb^{3+}), antimony V (Sb^{5+}), tin II (Sn^{2+}), and tin IV (Sn^{4+}). The **Group II** cations are precipitated out by reaction with 0.1 M H_2S at a low pH, normally around 0.5. The precipitates that are formed are also removed from the remaining mixture by centrifugation.

Group III cations are soluble in acidic H_2S but are insoluble in basic H_2S. **Group III** cations include aluminum (Al^{3+}), chromium III (Cr^{3+}), iron II (Fe^{2+}), iron III (Fe^{3+}), manganese II (Mn^{2+}), and zinc II (Zn^{2+}). These ions are precipitated out of the mixture by changing the pH of the previous solution (already saturated with 0.1 M H_2S) to a pH of 9 or greater.

All that is left at this point in the unknown solution should be the **Groups IV and V** cations. These are the alkali and alkaline earth metal cations and the ammonium cation. **Groups IV and V** cations include barium (Ba^{2+}), calcium (Ca^{2+}), and magnesium (Mg^{2+}) and potassium (K^+), sodium (Na^+), and ammonium (NH^{4+}), respectively. The alkaline earth metal cations Ba^{2+}, Ca^{2+}, and Sr^{2+} are precipitated in 0.2 M $(NH_4)_2CO_3$ solution at pH 10 and then removed from the others by centrifugation. The alkali metal cations K^+ and Na^+ are soluble in most solutions and thus cannot be separated by precipitation. Rather, their presence is confirmed by a colorimetric flame test. The ammonium ion is also soluble and may have been lost through earlier tests as ammonia. To test for the presence of ammonium, you should take a small amount of the **original** mixture and add NaOH until the mixture becomes basic. At that point, the smell of ammonia (NH_3) will confirm the presence of ammonium ion (NH^{4+}). Further confirmation can be made by heating the sample and holding a piece of moist red litmus paper at the mouth of the test tube. A color change to blue confirms the presence of ammonium ion in the solution.

The portion of qualitative analysis described above simply separates the groups from each other. To determine the presence of individual ions in each group, a variety of other reactions must be performed. Each reaction is specifically designed to react strictly with only one cation in each group, thereby confirming only its presence. As with the main group separations, the reactions with each cation rely on those chemical properties that are unique to that ion. The discussion of the separation and confirmation of those ions is quite extensive and is best broken up into several parts. A more detailed discussion of the confirmation tests for each cation in this experiment can be found in the background section of Experiment 24 (Cations).

Anions

As with cations, the anions in a mixture are similarly determined using confirmation reactions based on the unique chemical reactivity of the anion. The most common anions found are carbonate (CO_3^{2-}), sulfate and sulfite (SO_4^{2-}, SO_3^{2-}), phosphate (PO_4^{3-}), the halogens (Cl^-, Br^-, and I^-), nitrate and nitrite (NO_3^-, NO_2^-), acetate (CH_3COO^-), and chromate and dichromate (CrO_4^{2-}, $Cr_2O_7^{2-}$). A list of the reactions used to confirm the presence of each of these anions can be found in the background section of Experiment 23 (Anions).

The Experiment

As was mentioned in the introduction, qualitative analysis today is predominantly performed by instruments, so this procedure is somewhat historical in nature. As such, we have decided not to perform extensive qualitative analysis experiments, which often run for weeks to cover all of the separations described above. Rather, the experiment you will be performing will take only a single lab period but should give you experience in the types of reactions and techniques you would typically perform during a qualitative analysis scheme.

In your experiment, you will be investigating seven cations (Na^+, Mg^{2+}, Ni^{2+}, Cr^{3+}, Zn^{2+}, Ag^+, and Pb^{2+}) as well as four anions (NO_3^-, Cl^-, I^-, and SO_4^{2-}). You will use the reactivities of these ions to both separate and identify them. In the first part of this experiment you will investigate the cations and anions with the goal of gathering information about their reactivity and physical properties to use as a reference guide later for an unknown mixture of some of the same ions. It is extremely important that you take meticulous notes with respect to every aspect of a solution or precipitate's appearance. For example, don't just write down "a white precipitate"; comment on the structure of the precipitate, for example, flaky or crystalline or powdery, as there may be several white precipitates in the experiment.

Part I: Reactions of Cations with NaOH and NH₃:

Most hydroxide salts are insoluble and thus precipitate out of solution when formed by reaction. The precipitates that form from these reactions often have distinctive colors or textures, lending to their identification. Some hydroxide salts undergo further reaction to form oxides. For example:

$$2Ag^+(aq) + 2OH^-(aq) \Leftrightarrow 2AgOH(s) \Leftrightarrow Ag_2O(s) + H_2O(l)$$

If an excess of hydroxide is added, however, there are some hydroxide salts that will exhibit amphoteric character and redissolve, forming complex ions.

For example, aluminum in dilute hydroxide solution forms a white gelatinous powder, $Al(OH)_3$. But if more hydroxide is added, the aluminum hydroxide dissolves, and the soluble complex ion $Al(OH)_4^-$ forms instead.

The simplest way to determine the structure of a complex ion is to look it up in a reference book or your text. In this experiment, all of the complex ions containing OH^- will contain four OH^- ions. This generalization should be used strictly for this experiment. The overall charge on the complex ion will therefore be the charge on the metal cation minus the number of OH^- ions.

In addition to aluminum, you will also be looking at the reaction of hydroxide with aqueous ammonia, NH_3.

NH_3 solutions are basic enough to create insoluble hydroxides but do not supply an excess of OH^- ions in the solution. If you add a small amount of ammonia, insoluble hydroxides will precipitate.

Ammonia can also form complex ions with some metals. These complex ions are soluble and will generally have four NH_3 ligands bound to the metal, for example, $Zn(NH_3)_4^{2+}$. Silver is the only exception to the four ligand rule for this experiment; it has only two ammonia groups in its complex, $Ag(NH_3)_2^+$.

Part II: A Stability Sequence

A stability sequence is a sequence of salts and ions for a given cation ranked based on their stability from more stable to less stable. In the following series, the most stable salts are the ones farthest to the right. While you can perform reactions to make more stable salts (left to right), you cannot go in the opposite direction.

For Ag^+: $Ag_2O(s) < AgCl(s), Ag(NH_3)_2^+(aq) < AgI(s)$

For Pb^{2+}: $PbCl_2(s) < PbSO_4(s) < PbI_2(s) < Pb(OH)_2(s) < Pb(OH)_4^{2-}(aq)$ in xs OH^-

For Zn^{2+}: $Zn(OH)_2 < Zn(NH_3)_4^{2+}(aq) < Zn(OH)_4^{2-}(aq)$ in xs OH^-

For Ni^{2+}: $Ni(OH)_2(s) < Ni(NH_3)_4^{2+}(aq) < Ni(OH)_2(s)$ in xs OH^-

What these sequences indicate is that if, for example, you were to add aqueous sulfate ions to solid lead chloride, $PbSO_4$ would form. You could not, however, go in the other direction. But you could add iodide (I_2) to lead sulfate ($PbSO_4$) or lead chloride ($PbCl_2$) and get lead iodide (PbI_2) to form. Any species above shown as an ion is soluble in water. The neutral compounds are all solids and will precipitate.

Anion Precipitation Reactions

In this experiment, only two cations will form precipitates when combined with the anions we are using. Silver forms two precipitates, and lead forms three precipitates. You will need to test these anions (NO_3^{1-}, Cl^{1-}, I^{1-}, and SO_4^{2-}) with each of the two reactant cations and take careful notes to allow you to determine the identity of the anions in your unknown.

Procedure

SAFETY NOTES: There are several strong acids and bases used in this lab. Use great care when handling them and wash your hands immediately if you are splashed.

GENERAL INSTRUCTIONS: Each student should work independently for this lab.

Part I: Reactions of Cations with NaOH and NH₃

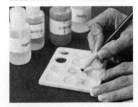

Obtain a well plate. Make sure it is clean and dry.

In separate wells, place 10 drops (~0.5 mL) of each of the cations except for Na^+. (All Na^+ salts are soluble, so we know that they will not react.)

Use a wax pencil to note the identity of each cation in the well to avoid confusion.

To each of the cation wells, add 1 drop of 6 M NaOH. Tap the plate gently to mix and carefully record observations of each cation.

Excess OH addition: Now add 10 more drops of 6 M NaOH to the same wells. Tap the plate gently to mix and carefully record observations of each cation.

Discard the used solutions in the well plate in the designated waste container and rinse well. Remove any excess water by shaking or Kimwipes.

Repeat the process above, adding NH₃ instead of NaOH to each of the cations. Tap the plate gently to mix and carefully record observations of each cation.

Discard the used solutions in the well plate in the designated waste container and rinse well. Remove any excess water by shaking or Kimwipes.

Part II: A Stability Sequence

Combine 1 mL of AgNO₃ and 1 mL of NaOH in a small test tube. Make and record your observations.

Centrifuge and discard the supernatant. Record all of your observations.

Add several eyedroppers full of aqueous NaCl solution to the test tube and mix well. Record all of your observations.

Centrifuge and discard the supernatant. Record all of your observations.

Add NH$_3$(*aq*) until the solid dissolves. Record all of your observations.

Add NaI until a reaction is seen. Record all of your observations.

Part III: Anion Precipitation Reactions

Place 10 drops of Ag$^+$ into four wells on your well plate.
Place 10 drops of Pb^{2}+ into another four wells on your well plate.

Add 1 drop of Cl$^-$ to one Ag$^+$ well and one Pb^{2+} well. Record your observations.
Add an additional 10 drops of Cl$^-$ to the same Ag$^+$ and Pb^{2+} wells. Record any changes.
Repeat the above process for the remaining three anions (I$^-$, SO$_4^{2-}$, and NO$_3^-$).

Part IV: Sodium Ions

Place a small amount of NaNO$_3$ solution on a watch glass.
Vaporize the NaNO$_3$ under a Bunsen Burner in the hood. An intense Yellow flame confirms the presence of sodium. See note in an earlier experiment about heating watch glasses.

Part V: Nitrate Ions

Place 20 drops of aqueous $NaNO_3$ in a test tube.

NOTE: If this were an unknown, you would need to remove any iodide from the sample first by precipitating it out using several drops of saturated Ag_2SO_4. Otherwise, you would get a false positive.

Slowly and very carefully add 20 drops of concentrated sulfuric acid.

Now add ~5 drops of Iron II Sulfate heptahydrate by allowing the drops to roll gently down the inside of the test tube. DO NOT MIX!

The formation of a smokey brown ring at the solution interface confirms the presence of Nitrate ion.

Part VI: The Unknown

Collect ~10 mL of one of the unknowns. Record the Code of the Unknown in your notebook.

Using the same tests and your recorded observations from above, determine the identity of the cation(s) and anion(s) present in your unknown.

Report Contents and Questions

The purpose should include a majority of the topics and techniques covered in the experiment. The procedure section should cite the lab manual and include any changes made to the procedure during lab work. The procedure section should also include the step-by-step procedure on how you determined the identity of the unknown.

The data section should include a table for Parts I and III listing the seven cations down the side and with five columns labeled: Initial Observation, OH^-, xs OH^-, NH_3, and Anion PPT. In the first column, record your initial observations, noting the color of each solution. In the remaining sections, describe the results upon mixing. The combinations that you did not mix in lab have no reaction. Be sure to include things such as colors and formulas of precipitates. DO NOT RECORD "NO CHANGE"; RECORD WHAT THE SOLUTION/MIXTURE LOOKS LIKE, EVEN IF IT IS CLEAR AND COLORLESS. For the anion precipitation column, indicate which anions form a precipitate and all other observations. All observations written in this table must match those made during lab and recorded in your lab notebook.

For Part II, make a table with the observations you recorded in lab after each step. For Parts IV and V, include all observations in a systematic fashion.

For Part VI (the identity of the unknown), include observations made during each step you took in lab, along with the code and identity of the cation(s) and anion(s) in your unknown.

The calculation section for this experiment will include balanced net ionic equations. Write the balanced net ionic equation for each reaction from Part I. If you observed a precipitate re-dissolve in excess OH^-, write two separate balanced equations: the first showing the formation of the precipitate from the metal ions and the hydroxide ion, and the second with the precipitates and additional hydroxide ions as the reactants. NH_3 was also reacted with all the cations except Na^+. If a precipitate formed and remained, a neutral hydroxide salt formed when the cation reacted with NH_3 and H_2O. If a precipitate was not observed or immediately redissolved, the cation reacted with NH_3 to form a complex ion. Formulas of any complex ions can be determined from the information provided in the background section of this experiment. ALSO write the balanced net ionic equations for the anion precipitation reactions you observed in Part III.

The conclusion section should include several detailed paragraphs with the following information: a paragraph explaining how a stability series works, including how the observations you made specifically illustrate the stability series for silver ions (this is from Part II); the identity of the ions in your unknown and how you reached these conclusions, including a discussion of the known reactions you did in Parts I through V and how they compare to the unknown reactions (be sure to include a description of both supporting and eliminating experiments that you conducted); and, finally, include all possible errors found in this experiment.

Answer the following questions:

1) Which cation forms a two ligand complex with NH_3?

2) Which anion does not form a precipitate with lead?

3) Write the balanced equation for the dissociation of the compound in question 2 in water.

Experiment 26
Laboratory Preparation

Name: _____ Date: _____

Instructor: _____ Sec. #: _____

Show all work for full credit.

1) Read the background, procedure, and report sections of the lab experiment carefully and develop a hypothesis of what information you expect to gain from the completion of the lab experiment.

2) Create any and all tables you might need to collect data during the experiment and then transfer those tables into your lab manual for use during the lab.

Experiment 26
Prelaboratory Assignment

Name: _____ Date: _____

Instructor: _____ Sec. #: _____

Show all work for full credit.

1) The qualitative analysis scheme depends on the relative solubilities of the salts of the cations with various anions. K_{sp}, the solubility product constant, describes the equilibrium reaction between the solid salt and the ions in solution:

$$M_xA_y(s) \leftrightarrow xM^{y+}(aq) + yA^{x-}(aq), \quad K_{sp} = [M^{y+}]^x[A^{x-}]^y$$

(M_xA_y does not appear in the expression because the pure solid is considered to be in its standard state with an activity of 1.)

The actual solubility may differ slightly from the calculated value, though, if the undissociated salt itself is slightly soluble and if there are intermediate ion-pair structures formed in solution. Nevertheless, the calculated value is still very useful when comparing solubility relationships of various ions and predicting whether a precipitate will form when two ionic solutions are mixed.

For the salt $BaSO_4$, $K_{sp} = 1.1 \times 10^{-10}$.
What is $[Ba^{2+}]$ for a saturated solution of $BaSO_4$?

What is the calculated solubility for $BaSO_4$?

Addition of a soluble sulfate, such as Na_2SO_4, will suppress the solubility of the Ba^{2+} ion because of the common ion effect.
What will $[Ba^{2+}]$ be in the presence of 0.15 M Na_2SO_4?

2) For the salt SrF_2, $K_{sp} = 4.3 \times 10^{-9}$.
What is $[Sr^{2+}]$ for a saturated solution of SrF_2?

What is the calculated solubility for SrF_2?

What will $[Sr^{2+}]$ be in the presence of 0.10 M NaF?

3) To estimate whether a salt will precipitate under a particular set of conditions, a **reaction quotient, Q_{sp},** can be calculated from given concentrations of the cation and anion using the equation:

$$Q_{sp} = [M^{y+}]^x[A^{x-}]^y$$

If Q_{sp} is less than or equal to K_{sp}, no precipitate will form, but if Q_{sp} is greater than K_{sp}, a precipitate will form, and the ion concentrations will decrease until they reach values that satisfy the K_{sp} expression.

In the case of metal hydroxides, the anion is the hydroxide ion, and its concentration depends on the pH of the solution. Whether a precipitate forms will therefore depend both on the K_{sp} of the hydroxide and on the pH.

For the reaction: $Mg(OH)_2 \Leftrightarrow Mg^{2+} + 2OH^-$, $K_{sp} = 5.6 \times 10^{-12}$.
You prepare a solution of 0.20 M Mg^{2+} and adjust the pH to 6.0.
What is the value of the reaction quotient, Q_{sp}, under these conditions? Will a precipitate form? What will be the equilibrium concentration of the Mg^{2+} ion?

4) Insoluble salts can sometimes be brought into solution by the formation of a soluble complex ion. For example, AgCl(s) can be dissolved in ammonia solutions by the formation of the soluble $Ag(NH_3)_2^+$ complex ion. The ammonia competes with the Cl^- ion for reaction with Ag^+ as follows:

$$AgCl(s) \leftrightarrow Ag^+(aq) + Cl^-(aq) \qquad K_{sp} = [Ag^+][Cl^-] = 1.8 \times 10^{-10}$$

$$Ag^+(aq) + 2NH_3(aq) \leftrightarrow Ag(NH_3)_2^+(aq) \qquad K_{formation} = \frac{[Ag(NH_3)_2^+]}{[Ag^+][NH_3]^2} = 1.6 \times 10^7$$

Sum: $AgCl(s) + 2NH_3(aq) \leftrightarrow Ag(NH_3)_2^+(aq) + Cl^-(aq)$

$$K_{net} = K_{formation} \times K_{sp} = \frac{[Ag(NH_3)_2^+][Cl^-]}{[NH_3]^2} = 2.9 \times 10^{-3}$$

Assuming all the Cl^- comes from the dissociation of AgCl, then $[Ag(NH_3)_2^+] = [Cl^-]$, and $[Ag(NH_3)_2^+]^2 = [NH_3]^2 \times 2.9 \times 10^{-3}$ or $[Ag(NH_3)_2^+] = [NH_3] \times \sqrt{2.9 \times 10^{-3}}$.

What is the maximum $[Ag(NH_3)_2^+]$ that can be obtained by dissolving AgCl in 1.5 M ammonia?

5) For another silver salt, $AgBrO_3$, $K_{sp} = 5.4 \times 10^{-5}$.

What is K_{net} for the reaction $AgBrO_3 + 2NH_3 \leftrightarrow Ag(NH_3)_2^+ + BrO_3^-$?

What is the maximum $[Ag(NH_3)_2^+]$ that can be obtained by dissolving $AgBrO_3$ in 4.1 M ammonia?

Experiment 27

Whodunit? A Forensic Investigation

Introduction

Forensic chemistry has captured the interest of many nonchemists because of the very popular CSI television shows. As one of the most visible and readily understandable applications of chemistry techniques, forensic methods are often more interesting because of what they can prove or disprove rather than for the chemistry behind them. But as we are in a chemistry lab, and therefore should be learning both the technique and the chemistry behind it, the focus of this lab will be on developing new skills and chemical knowledge. But... we don't want to take all the fun out of it, so we included a mystery to solve. You will investigate a murder using three different methodologies: simple fiber burn test, iodine fuming to develop fingerprint evidence, and thin-layer chromatography (TLC) to determine ink similarities. Each of these techniques is currently used in forensic laboratories throughout the United States and around the world.

You will receive an evidence bag and work with other investigators to determine if either of two suspects may be the perpetrator. The bag contains a threatening note and some fibers collected at the crime scene. You will also have access to the case file and should read through the detective's notes to understand the relevance of each piece of evidence. Make sure you bring to class any of the case file materials you will need during your investigation (lab period).

Background

In this experiment, you will act as a criminalistics technician, analyzing evidence collected at the scene of a hypothetical homicide. You will use three real-life forensic techniques based entirely on chemical principles to determine the probability of guilt of two suspects. You will use thin-layer chromatography to compare the inks found on a threatening note left at the crime scene to pens found in the possession of the two suspects. Using the same note, you will use iodine fuming to develop fingerprints to compare with those of your two suspects. Finally, you will attempt to determine the fiber content of a piece of torn fabric found clenched in the hand of the victim.

Ink Analysis

The analysis of inks and dyes is an important technique for a document examiner. Such a technique is used daily to detect the forgery of documents and the falsification of signatures. Ink analysis has even been used to provide an estimate of the date at which a particular document was signed. Ink analysis can be employed to test for the adulteration of numbers that increase the amount of money being charged, to observe whether ink used in two parts of a document came from the same pen, or even to compare the writing used in a threatening letter with pens confiscated from a suspect's office, which is what you will do in this experiment.

Thin-Layer Chromatography

Liquid inks used in ballpoint pens contain a mixture of substances, including dyes, stabilizers, solvents, and resinous binders. Some manufacturers also add fluorescent markers or heavy metals to aid in the identification of their particular product. The preferred technique for ink analysis is thin-layer chromatography (TLC). TLC is one member of a family of analytical techniques utilized for separating mixtures, such as liquid inks. TLC is usually performed on plates that have been uniformly coated by silica, alumina, cellulose, or some other type of adsorbent.

Ink to be sampled by TLC will be punched out of paper evidence and extracted using pyridine. Once this is accomplished, the sample is spotted onto the TLC plate by means of a glass capillary used as a transfer pipette. The plate is then placed in a glass tank containing a layer of solvent at the bottom. As the solvent moves up the plate by capillary action, the components of the sample are separated on the basis of each one's unique combination of attractions for the coating (stationary phase) and the solvent (mobile phase). Generally, a TLC is run with the unknown sample and samples of the suspected "knowns" so that a direct visual comparison is easily made. However, quantitative comparisons are also required for definite identification.

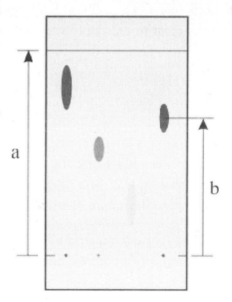

The retention factor (R_f) is a numerical characteristic calculated for each band in a chromatogram. Comparisons of standard R_f values to those of unknown compounds are used to determine the identity of the unknown components. The retention factor (R_f) is determined by measuring the distance traveled by an individual band (b) on a chromatogram and dividing that distance by the distance traveled by the solvent front (a). Because it is a ratio, an R_f value does not have any units. Notice that the bands can be "smeared" or stretched. It is important that you select the top, bottom, or middle of the band to measure and be consistent from measurement to measurement. Another important detail is that although the solvent front in this diagram is shown as a straight line, quite often this is not the case. If the solvent line is "wavy" or on a slant, measure the solvent distance directly above each band to get an accurate R_f value. The R_f values that should be compared when determining the identity of the ink found on the evidence.

Fingerprint Analysis

It is well accepted that fingerprints are unique to each individual and can thus be used for identification. In evidence, prints may not be visible. If the prints are not visible, electronic, physical, and—yes—chemical processing may allow the print to be visualized. Once visualized, the technician must then provide a comparison report to determine if the prints are a match to a suspect's prints. Print comparison is now run digitally, but unlike depictions on CSI shows, computer searches produce only some probable suspects. The final identification and comparison is generally done by hand. For the fingerprint analysis, you will use iodine fuming to develop any latent prints present on the note. As part of the case file, you will have reference print cards of the two suspects for comparison. Other techniques, such as super glue fumes, ninhydrin, and silver nitrate are additional options.

The process used to develop the fingerprints is called iodine fuming. The development of latent prints with iodine fumes is not a chemical process; it is a physical one. In the iodine fuming process, the natural body fats and oils found in a latent print temporarily absorb the iodine vapors. This results in a change in color, from clear to a dark brown, until the effect fades with time.

Fiber Analysis

The last forensic technique you will use in this experiment, and perhaps the easiest to perform, is the fiber analysis. A piece of cloth has been retrieved from the crime scene and may be from an article of the perpetrator's clothing. To connect the criminal to the crime scene, the type of cloth must be identified. Normal fiber analysis would involve the use of a microscope or possibly a gas chromatograph, but as you have neither of these at your disposal, you will use the burn test. This is a simple and inexpensive way to determine the nature of a fiber.

The burn test is used to determine if the fabric is a natural fiber, a human-made fiber, or a blend of natural and human-made fibers. It takes practice with the burn test to determine the exact fiber content; however, an inexperienced person can still determine the difference between many fibers to narrow the choices down to natural or human-made fibers. The advantage of this test for forensic purposes is the short amount of time it takes to complete it.

Identifying the Fibers

Cotton is a plant fiber. When ignited, it burns with a steady flame and smells like burning leaves. The ash left is easily crumbled. Small samples of burning cotton can be blown out as you would a candle.

Linen is also a plant fiber but different from cotton in that the individual plant fibers that make up the yarn are long where cotton fibers are short. Linen takes longer to ignite. The fabric closest to the ash is very brittle. Linen is easily extinguished by blowing on it as you would a candle.

Silk is a protein fiber and usually burns readily, though not necessarily with a steady flame, and it smells like burning hair. The ash is easily crumbled. Silk samples are not as easily extinguished as cotton or linen.

Wool is also a protein fiber but is harder to ignite than silk, as the individual "hair" fibers are shorter than silk, and the weave of the fabric is generally looser than with silk. The flame is steady but is more difficult to keep burning. The smell of burning wool is like burning hair.

Human-Made Fibers

Acetate is made from cellulose (wood fibers), which is known as cellulose acetate. Acetate burns readily with a flickering flame that cannot be easily extinguished. The burning cellulose drips and leaves a hard ash. The smell is similar to burning wood chips.

Acrylic, called acrylonitrile, is made from natural gas and petroleum. Acrylics burn readily due to the fiber content and the lofty, air-filled pockets. A match or cigarette dropped on an acrylic blanket can ignite the fabric, which will burn rapidly unless extinguished. The ash is hard. The smell is acrid or harsh.

Nylon is a polyamide made from petroleum. Nylon melts and then burns rapidly if the flame remains on the melted fiber. If you can keep the flame on the melting nylon, it smells like burning plastic.

Polyester is a polymer produced from coal, air, water, and petroleum products. Polyester melts and burns at the same time, and the melting, burning ash can bond quickly to any surface it drips on, including skin. The smoke from polyester is black with a sweetish smell, and the extinguished ash is hard.

Rayon is a regenerated cellulose fiber and is almost pure cellulose. Rayon burns rapidly and leaves only a slight ash. The burning smell is similar to burning leaves.

Blends consist of two or more fibers and, ideally, are supposed to take on the characteristics of each fiber in the blend. The burning test can be used, but the fabric content will be an assumption.

339

There are, of course, many other types of fibers that are not listed here. For the purposes of this lab experiment and investigation, we will assume the library provided will be adequate for proper identification of any unknown fibers found.

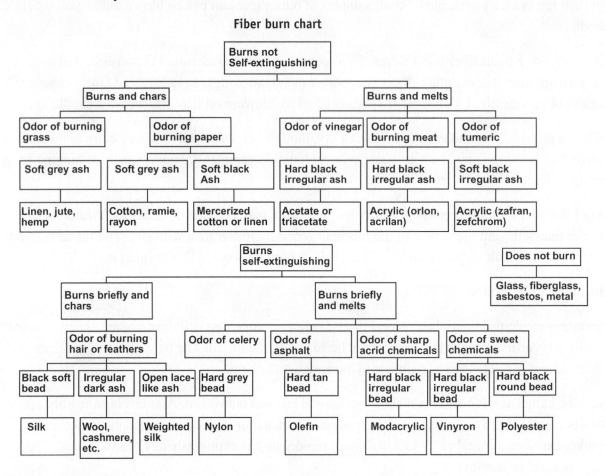

Fiber burn chart

Another note on the techniques you will be performing: burn tests and iodine fuming are seldom used in a research lab, but TLC is a very commonly used separation technique with many applications beyond forensics. When purifying a protein to study, a biochemist may use TLC to separate one protein from another or to simply confirm the presence of the protein he or she needs. The principles of TLC also can be seen in related methodologies such as DNA restriction fragment length polymorphism (RFLP) identification, commonly called DNA fingerprinting. The separation of DNA for RFLP is done by means of a TLC process. The stationary phase in a DNA RFLP experiment is a gel as opposed to paper or cellulose, and the migration and separation of the DNA is assisted by electrical current, but the resulting chromatogram is very similar in looks and information.

Hopefully, at the completion of these techniques, you will be able to suggest that the district attorney prosecute a suspect or continue searching! Now, read your case file and get to work!

CASE FILE

Detective's Narrative

On the morning of January 2, 2015, I was called to the Department of Chemistry and Biochemistry at the University to investigate the death of Justin Trueblood, a second-year graduate student in the Biochemistry Division. At 8:13 a.m., Trueblood's body was found by the maintenance personnel lying on the floor of the second floor men's restroom in front of the stalls. It was obvious from bruising that the victim had been strangled. Autopsy is pending. Estimated time of death was 3:00 a.m. Restroom was in general disarray indicative of a struggle. Using the building's key card entrance system records, I was able to determine that only two other individuals (John Badman and Dementia Jones) had been keyed into the building around the estimated time of death. Both individuals were still in the building upon my arrival. Collection of fingerprints and personal materials was made at that time.

During questioning, it was revealed that the suspects and the victim all worked in the same third floor laboratory and that Justin had previously had a personal relationship with Dementia. But according to Dementia, the relationship ended several months ago, and she is now engaged to John Badman. According to Dementia, there was no trouble between the three of them, and they were working together when Justin left around 2:00 a.m. and did not return to the lab. She assumed that he had gone home for the night. According to Dementia, John was with her in the lab all night long.

When questioned, John confirmed that Justin had left the lab about 2:00 a.m. and had not returned. When asked about his whereabouts during the course of the night, John also said he was present in the lab all night. I asked him about restroom breaks, and he did admit to leaving the lab once around 5:00 a.m. but claimed to have used the fourth floor restroom, not the second.

A lab coat containing several recent tears (edges were frayed but not dirty) was found in the laboratory. The coat was unclaimed by either suspect, who stated they did not know who it belonged to and that it had been there for a long time.

B.F. #65773

Crime scene photo

Procedure

SAFETY NOTES: The TLC solvent should be used only in the hood. Do not inhale the iodine fumes, as they are harmful to mucous membranes. The iodine crystals should be kept strictly in the hood. Make sure the fingerprint papers are fully "aired" before removing them from the hood area. Technicians must wear gloves during all procedures to prevent contamination of the evidence.

GENERAL INSTRUCTIONS: Each bench will work as a forensics team. An evidence bag is provided to each bench. Be sure to note the identification codes on all pieces of evidence.

Part I: Identification of Ink Type Using TLC

Every team member should put on gloves before handling any evidence.

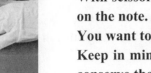

Carefully remove the "threatening note" from its evidence bag. Record your observations about the note in your lab notebook.

With scissors or a scalpel, carefully cut out two small slivers of the writing on the note.
You want to have mostly ink on these samples.
Keep in mind that you will be testing this note for fingerprints as well, so conserve the evidence as much as possible.

On another separate piece of paper, make test writings with the pens obtained from the suspects.
Cut slivers of ink from each suspect note as above.

Place each set of ink sample slivers in a different well in a well plate.
Add 1 to 2 drops of methanol to each well.
Use as little methanol as possible to extract the ink so that it is not too dilute.
Let the plate sit for 1 to 2 min. Shake gently, if necessary, to remove the ink from the papers.

Prepare a TLC plate by lightly drawing a straight line with a pencil across each plate 1cm from the bottom.

Spot the plate using capillary tubes by placing them into the extracted samples and drawing out a small amount of solution. You should have a total of four spots, the three pens collected as references and your unknown ink from the note.
Apply a small spot, let dry, and repeat until a very small clearly visible spot appears on the plate.
Mark the identity of each spot in pencil at the top of the plates. Be careful to not let the spot become too big, or your experiment will be ruined.

Prepare a TLC tank by adding ~25 mL of chromatography solution to a 400 mL Beaker.

Place the TLC plate in the TLC Tank in the hood. Cover the tank with Parafilm or a watch glass.

Watch the solvent front move slowly up the plates. When the solvent is $\sim\frac{3}{4}$ of the way up the plate, remove it from the tank.

Mark the solvent line with a pencil. Place the plates in the hood to dry.

Take the dry plates out of the hood and examine them using both visible and UV light. Mark the solvent front and any spots with a pencil. Calculate *Rf* value for each band.

Part II: Developing Fingerprints

Take the rest of the suspect threatening note and place a paper clip with a piece of string attached to it onto the top center of the note.

Tape the string and note to the inside bottom of a 1 L beaker so that when the beaker is turned upsidedown, the note will be suspended in the center.

Take the beaker and note to the hood. Set the inverted beaker over a watchglass containing iodine crystals.

Allow the fingerprints to develop for approximately 3 to 5 min.

Remove the note from the beaker and wave the note around to air out any unfixed iodine fumes.

Take the note back to your lab bench and, using the reference fingerprint cards from the background for comparison, identify (if possible) the suspect.

Part III: Fiber Analysis

Carefully remove the fiber evidence from the evidence bag. Record your observations.

Cut a small portion of the fiber from the whole sample.

Holding the fabric with tweezers, not your fingers, over a watch glass, burn the fibers using a bunsen burner. Make and Record your observations. Repeat if necessary to draw a conclusion as to the type of fiber (see background).

Report Contents and Questions

Although you work as a team for the experiment, each technician should complete his or her own report.

The purpose should include the various topics covered throughout the experiment and the criteria that will be used to determine the success of the investigation. The procedure section must cite the lab manual and include any details about where your investigation took you with respect to the evidence. Also, be sure to note any changes that are made during lab to the procedure given in your laboratory notebook.

The data section should include a completed copy of the:

1) Crime Scene Unit Property Report

2) Property Receipt

3) Request for Comparison

4) Latent Print Report

Also include any and all observations that you recorded during the lab.

The calculations section should include an example calculation of how the R_f values for your TLC plates were determined.

The conclusion section should be several well-thought-out paragraphs containing a complete description of the results of each forensic test and your analysis of those results. A discussion of possible errors in the testing is required.

Answer the following questions:

1) What are the criteria for determining the guilt or innocence of a suspect? Does the evidence you have collected meet that criteria?

2) As a scientist analyzing data, should you have an agenda?

3) Iodine fuming is used in this lab to develop latent fingerprints. What are latent fingerprints? What are some other methods by which latent prints can be developed? What are some of the advantages to using iodine crystals for the development? What are some of the disadvantages?

Experiment 27
Laboratory Preparation

Name: _____ Date: _____

Instructor: _____ Sec. #: _____

Show all work for full credit.

1) Read the background, procedure, and report sections of the lab experiment carefully and develop a hypothesis of what information you expect to gain from the completion of the lab experiment.

2) Create any and all tables you might need to collect data during the experiment and then transfer those tables into your lab manual for use during the lab.

Experiment 27
Prelaboratory Assignment

Name: _____ Date: _____

Instructor: _____ Sec. #: _____

Show all work for full credit.

1) Classify the following prints as to "arch," "tented arch," "loop," or "whorl."

_____   _____   _____   _____

Fingerprint images are from *NIST* and *Bergen County Technical Schools* web pages. The website is www.nist.gov/itl/iad/ig/fingerprint.cfm and sites.bergen.org/forensic/fingerprint.htm

2) The figure to the right shows a TLC plate in which an ink sample to be identified has been separated into three different pigment components.

Determine the *Rf* value for each of the three pigments. (Measure the distance from the origin to the center of the spot.)

3) Below are TLC plates for ink from six different reference sources.

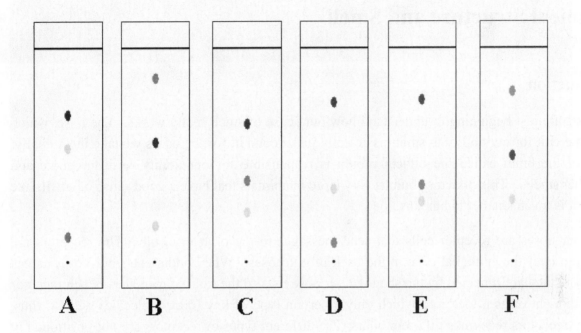

Identify the ink source that corresponds to the unknown sample in problem 2. Show calculations to support your answer.

Experiment 28
Chemical Structure and Smell

Introduction

Scientists are just beginning to understand how our sense of smell really works. The nose, which is given credit for our ability to smell, is actually just a conduit for the odors we take in. Actually, a part of the brain called the olfactory bulb is responsible for our ability to differentiate and remember smells. This olfactory bulb is very large in animals that have a good sense of smell, like dogs, but is much smaller in humans.

Inside our noses are receptor cells that send messages to the olfactory bulb. The receptors are spread out over the epithelial cells in the back of our noses. When odors enter our noses as gas molecules, they are funneled by the shape of our nose toward the field of receptors. Each receptor can be thought of as a lock into which only a certain type of key (odor molecule) will fit; thus, different receptors recognize different odors. The different types of receptors are not positioned in groups, rather, they are spread throughout the field of receptors so that when they are activated by an odor, the message sent to the olfactory bulb is that a certain combination of receptors has been activated. The bulb then translates this to mean "something smells fishy" or "chocolate chip cookies," depending on the particular combination of receptors. The different combinations of receptors being activated in different positions are what give us the ability to identify thousands of different smells.

The "lock and key" mechanism of the receptors is of interest to us in the general chemistry lab. If receptors differentiate smells by the shape of the molecules, maybe we can too. In the lab today, you will be presented with a number of different chemicals that are responsible for very different smells. The molecules are all organic in nature and vary in size, shape, and the functional groups they contain. It will be your job to first identify the molecules strictly by their smell in comparison to the descriptions in the background section. You will then match your identified molecules to one of the molecular models provided.

Background

In the early 1970s, John Amoore published an article in the highly acclaimed journal *Nature*. He proposed the stereochemical theory of odor, which suggested that there are only five types of receptors of smell for every substance, a gross oversimplification, but we'll use it for now. A table listing Amoore's proposed receptors, the shapes of the molecules they would interact with, and the odor associated with them is given below:

Receptor	Molecule Shape	Odor
Camphoraceous	Football	Mothball
Musky	Necklace	Musky
Peppermint	Wedge	Peppermint
Floral	Tad-pole	Flowery
Ethereal	Long, thin ether molecule	Ether

Today, it is known that there are more than 1000 smell receptors. However, Amoore was not all wrong, as we will see how distinct molecular shapes, as well as alterations to a molecule's structure, affect the perception of a particular chemical's odor. For example, it was only recently that perfume chemists figured out that the addition of an organic hydrocarbon to a molecule is one way to increase a fragrance's potency.

The table below lists by common name and (chemical name) information about 12 well-known molecules and their respective smells. It will be your job in this experiment to use this information to identify a selection of 6 of these molecules presented as unknowns in both chemical form (from smell) and as molecular models.

Molecule	Formula	Where found?	Category	Functional groups	FYI
Acetic acid (Ethanoic acid)	$C_2H_4O_2$	Sour Wine	Acid	Carboxylic acid	$T_m = 16.6\ °C$ $T_b = 255.8\ °C$
Diacetyl (2,3-butanedione)	$C_4H_6O_2$	Butter, cheese	Fatty	2 Ketones	$T_m = -2.4°C$ $T_b = 88.0°C$
Isoamyl acetate (1-butyl-3-methyl-acetate)	$C_7H_{14}O_2$	Pear	Fruity	Ester	$T_m = -78.0\ °C$ $T_b = 142.0\ °C$
beta-Phenylethyl alcohol (2-phenylethanol)	$C_8H_{10}O$	Rose	Floral	Alcohol and phenyl	$T_m = -27.0\ °C$ $T_b = 218.2\ °C$
Isoamyl propionate (1-butyl-3-methyl propanoate)	$C_8H_{16}O_2$	Pineapple	Fruity	Ester	$T_m = N/A$ $T_b = 156.0\ °C$
Cinnamaldehyde ((E)-3-phenyl-2-propenal)	C_9H_8O	Cinnamon	Spices	Phenyl, alkene, and aldehyde	$T_m = -7.5\ °C$ $T_b = 246.0\ °C$
l-Carvone (2-methyl-5-(prop-1-en-2-yl) cyclohex-2-enone)	$C_{10}H_{14}O$	Spearmint	Spices	Alkene and ketone	$T_m = 88.9\ °C$ $T_b = 230.0\ °C$

Trimethylamine (N,N-dimethyl-methanamine)	C_3H_9N	Rotten fish, ammonia like	Animal	3 Amines	$T_m = -117.1\ °C$ $T_b = 2.9\ °C$
Pyridine (Azabenzene)	C_5H_5N	Coal tar	Burnt	N/A	$T_m = -41.6\ °C$ $T_b = 115.2\ °C$
Putrescine (1,4-diaminobutane)	$C_4H_{12}N_2$	Rotting flesh	Animal	2 Amines	$T_m = 27.0\ °C$ $T_b = 158.0\ °C$
t-Butyl Mercaptan 2-methyl-2-propanethiol	$C_4H_{10}S$	Skunks	Animal	Thiol	$T_m = -0.5\ °C$ $T_b = 64.0\ °C$
Allicin (2-propene-1-sulfinothioic acid S-2-propenyl ester)	$C_6H_{10}OS_2$	Garlic	Spice	Alkane and thiol	$T_m = <25.0\ °C$ $T_b = N/A$

Procedure

SAFETY NOTES: Never inhale chemicals directly! Waft the air above the chemical toward you. Wafting is waving your hand above the fumes and pulling them toward you. Slowly smell the odor.

Part I: Identifying and Classifying the Smells

Visit each of the six "smell" stations located at the various benches throughout the lab.
Using a wafting motion, smell each of the unknown samples and record your observations in your lab notebook.

Using the information given in the background section, determine the identity of each of the unknowns.

Part II: Determining and Comparing Chemical Structures

Using the chemical name given for each of your now identified unknowns, identify the corresponding chemical structure from the models displayed at the front of the lab.
Draw the structure in your lab notebook.
NOTE: If a model of your chemical does not appear to be present, you may want to reconsider the identification you made in Part 1.

Report Contents and Questions

The purpose should include the topics and techniques covered in the experiment as well as a statement of the criteria for success. The procedure section should cite the lab manual and include any changes made to the procedure during lab work. For this experiment, the procedure section should also include the steps that you took to determine the identity of the molecules; this should be complete enough so that someone else could read the procedure, repeat the experiment, and obtain the same results.

The data section should have the following for each molecule: **(a)** the number associated with the unknown smell and the letter of the matching molecular model; **(b)** the observations made during lab of the molecule's smell; **(c)** the name of the molecule; **(d)** the molecular formula of the compound; **(e)** the Lewis dot structure of the molecule, drawn in ink; **(f)** the molecular shape of the molecule; and **(g)** a list of any functional groups in the molecule. Be sure that all observations match those in your lab notebook. If you use any outside sources of information, be sure to cite them.

There are no calculations or graphs for this experiment.

The conclusion for this experiment should be several paragraphs long and include a discussion of the identity of each of the unknown molecules. This should be supported by your observations and background knowledge. Be sure to include in your discussion the information in your data section such as molecular shape, hybridization, functional groups, and anything else that allowed you to correctly identify the molecule. BE SURE TO CLEARLY STATE THE IDENTITY OF EACH UNKNOWN. Also discuss how the smell is related to molecular structure. Discuss similarities in smell (sweet, sour, fruity, rotten, etc.) with respect to organic functional groups. Include any possible errors in the experiment or identification of the unknowns.

Answer the following questions:

1) Give an example of two molecules with the same molecular formula but different smells. Explain why this difference occurs.

2) Long time use of cigarettes often leads to a loss or change of the sense of taste. Based on this experiment, describe how receptors for taste might function.

Experiment 28
Prelaboratory Assignment

Name: _____ Date: _____

Instructor: _____ Sec. #: _____

Show all work for full credit.

1. For each molecule listed in the table in the background section: **(a)** draw a correct Lewis Structure, and **(b)** circle and identify by name all of the functional groups in the molecule. *Copy your answers into your lab notebook for use during lab.*

2) Odor receptors detect molecules according to their size, shape, and the presence of certain functional groups. For the unknowns in your laboratory, see if you can first identify the presence of the following functional groups in their structures.

Which of the following have an alkene functional group?
acetic acid   cinnamylaldehyde   allicin   benzene   ethanol

Do the following have an aromatic functional group?
isoamyl acetate   cinnamaldehyde   diacetyl   isoamyl propionate

Do the following have a ketone functional group?
l-carvone   pyridine   putrescene   isoamyl propionate

Do the following have an aldehyde functional group?
t-butyl mercaptan   acetic acid   trimethylamine   pyridine

Do the following have a carboxyl functional group?
acetic acid   cinnalaldehyde   pyridine   isoamyl propionate

Do the following have an ester functional group?
diacetyl   allicin   l-carvone   t-butyl mercaptan

Do the following have an amine functional group?
benzeneethanol   trimethylamine   isoamyl acetate   putrescene

Do the following have an alcohol functional group?
acetic acid   benzeneethanol   cinnamylaldehyde   allicin

Experiment 29
Proteins, Carbohydrates, and Fats: Analysis of a Peanut

Introduction

Biochemistry is a specialty within the study of chemistry, much the same way cardiology is a specialty within the study of medicine. While still concerned with chemical reactivity, the focus of biochemistry is the molecules and chemical reactions that are found in living organisms. The most common molecules studied in biochemistry are proteins, carbohydrates (sugars), fats (more properly called lipids), and nucleic acids. Although the manipulation of nucleic acids is beyond the scope of this course, the other molecules can be identified and in some cases quantified using the unique chemical characteristics of each molecule.

Remember the old saying "You are what you eat"? In the case of the peanut, this really is true. The human body is made up of proteins (muscle) and fats, and we use carbohydrates as fuel (energy) to run all of our metabolic systems. The peanut, though small, contains a large quantity by weight of each of these biochemical molecules. Peanuts even contain enough protein to be recommended as a meat substitute for those individuals who eat a vegetarian diet! The table below lists the approximate percentages by weight of some of the components of a peanut kernel:

Constituent	Weight (%)
Water	3
Lipid	50
Protein	25.8
Carbohydrate	18.7
Inorganic salts (calcium, iron)	2.5

Data from: http://www.nal.usda.gov/fnic/foodcomp/cgi-bin/list_nut_edit.pl

Background

The peanut was first brought to the attention of the United States as a good source of fats, proteins, vitamins, and minerals by George Washington Carver in the early 1920s. Since then, peanut butter has become a staple food item in most American kitchens.

In today's experiment, you will identify several biochemical constituents of the kernel of a peanut and isolate an important lipid, peanut oil.

Isolating and Analyzing Lipids

Most of the weight (~50%) of a peanut is found in its lipid content. The percentages of the different fatty acid residues found in the triglycerides of peanut oil are ~7.5% palmitic, ~4% stearic, ~60% oleic, and ~19.5% linoleic. Separating these fatty acid constituents from the rest of the peanut can be accomplished by extraction with an organic solvent, methylene chloride, CH_2Cl_2. The peanut oil, which is predominantly hydrocarbon, is soluble in the solvent, whereas carbohydrates and proteins because of their polar and amphoteric nature are not. By grinding the peanut kernels into a paste and extracting the oil with methylene chloride and then evaporating the solvent, you can isolate the peanut oil and remove it from the other constituents. However, separating the oil into the component fatty acids requires gas chromatography, which is beyond the scope of this course.

Triglycerides (also called triacylglycerides) are the most common members of the lipid family. Triglycerides are esters in which the alcohol portion of the molecule is always a glycerol. The acid portion of the molecule is generally a long-chain acid, called a fatty acid.

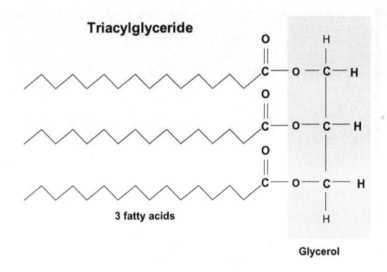

Triacylglyceride

3 fatty acids

Glycerol

Each R group in the triglyceride may be a different fatty acid such as the ones mentioned above. Saturated fatty acids contain the maximum number of hydrogens atoms per carbon in their chain. Unsaturated or polyunsaturated fatty acids contain one or more double bonds, respectively. The primary points of reaction for a triglyceride are at the ester and at the alkene (double bond, if it exists) functional groups. In this experiment, you will use the acrolein test to obtain evidence that glycerol is the alcohol in the triglyceride, and you will use alkene reactivity with potassium permanganate ($KMnO_4$) to test for unsaturation.

Acrolein Test

The acrolein test is performed by heating a triglyceride in a solution containing potassium hydrogen sulfate ($KHSO_4$) to form glycerol. The glycerol then further reacts with the $KHSO_4$ to produce acrolein, an unsaturated aldehyde with a sharp, penetrating odor.

| Triglyceride | Glycerol | | Acrolein |

Test for Unsaturated Fatty Acids

Potassium permanganate reacts with alkenes but not alkanes or aromatics. Evidence for a positive reaction, and the presence of alkene functional groups is provided by the disappearance of the purple $KMnO_4$ color and the formation of a brown manganese oxide (MnO_2) solution.

$$3R_2C{=}CR_2 + 2KMnO_4 + 4\,H_2O \rightarrow 3R_2(OH)C{-}C(OH)R_2 + 2MnO_2 + 2KOH$$

   an alkene   a glycol   (brown)

Testing for Carbohydrates

Carbohydrates are the primary energy source used by our bodies. Excess carbohydrates are converted to fats and stored. This is why consuming large quantities of sugar can make you fat and also why so-called energy bars often contain >50% sugar.

All monosaccharides and disaccharides, and some polysaccharides, are soluble in water but insoluble in organic solvents. Therefore, once the lipid content of the peanut kernel is removed, the carbohydrate (saccharide) content will remain in the meal of the peanut. All monosaccharides and many disaccharides are capable of reducing weak oxidizing agents such as the Cu^{2+} in Fehling's reagent.

Fehling's Reagent

These saccharides are referred to as reducing sugars. To be considered a reducing sugar, the carbohydrate must contain an aldehyde or hemiacetal (see below) functional group. It is only the acyclic or straight chain forms of sugars that are capable of being oxidized by Fehling's reagent. The acyclic form is less than 1% of the total sugar forms found in nature. Hydrolysis can be used to cleave the acetal bonds present in disaccharides and polysaccharides to produce an increased concentration of monosaccharides and thereby increase the response to the reagent.

$$R-\overset{\displaystyle O}{\underset{\displaystyle H}{C}} \; + \; 4OH^- \; + \; 2\,Cu^{2+} \longrightarrow$$

$$R-\overset{\displaystyle O}{\underset{\displaystyle O-H}{C}} \; + \; Cu_2O \; + \; 2\;\overset{H}{\underset{H}{O}}$$

Fehling's Test

Testing for Protein

Proteins are natural polymers that perform many essential functions in living systems. They carry oxygen from our lungs to our brains and other parts of the body, catalyze reactions, control pH, lift objects (as muscle), and transmit nerve impulses that allow you to move, breathe, and think. Protein obtained from the food we eat can provide both energy and essential amino acids. Proteins themselves are huge molecules consisting of long polymerized chains of individual amino acids. Their molecular weights are often >10,000 g/mol and have been found to be as great as 1,000,000 g/mol. The order in which the amino acids are bound together by amide (peptide) bonds is called the protein's primary structure. Special three-dimensional structures found in proteins such as beta-sheets or alpha-helices are referred to as secondary structures. The overall structure of the protein is called its tertiary structure.

Two of the amino acids very commonly found in proteins are tyrosine and tryptophan. Both amino acids contain aromatic rings. A test called the xanthoproteic test can be used to detect the presence of these two amino acids by reacting them with concentrated nitric acid (HNO_3). The nitric acid replaces some of the hydrogens in the reactive aromatic rings with nitro groups; treatment with a strong base then produces a yellow-orange color in solution.

$$\text{H}_3\overset{+}{\text{N}}-\overset{\text{H}}{\underset{\text{CH}_2}{\text{C}}}-\overset{\text{O}}{\overset{\|}{\text{C}}}-\text{OH}$$

Nitrated Tyrosine and Tryptophan

Since tyrosine and tryptophan are found in virtually all proteins, a positive xanthoproteic test is generally accepted as indicating the presence of protein.

Today's Experiment

Each of you will receive a peanut to be analyzed for lipid, carbohydrate, and protein content. The lipid content will be examined quantitatively by weighing the amount of oil extracted and determining the mass percent of oil in the peanut kernels examined. The lipid content will be confirmed qualitatively using potassium permanganate ($KMnO_4$) which reacts with unsaturated (alkene-containing) fatty acids and produces a brown color in solution. You will further use the acrolein test to test the lipids for the presence of triglycerides. The carbohydrate content of the peanut's kernel will be confirmed using Fehling's reagent. Finally, the presence of protein in the peanut kernel will be confirmed using the xanthoproteic test.

Procedure

SAFETY POINTS: Work with dichloromethane/methylene chloride only in the hood. Wear gloves when handling concentrated HNO_3 to avoid performing a xanthoproteic test on your skin or fingernails, which are both proteins.

Part I: Extraction of Peanut Oil

Separate the shells and kernels of two large peanuts. Remove the reddish skin.

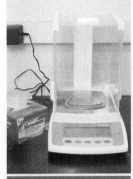

Weigh the four kernels to 0.0001 g and record the mass in your notebook.

Using a mortar and pestle, grind the peanut into a paste. Record your observations of the paste in your lab notebook.

In the hood, transfer the paste to an evaporating dish, rinsing the mortar with a few milliliters of methylene chloride.
Add 10 additional milliliters of methylene chloride to the evaporating dish. Continue to grind the peanut for a few minutes. When done, cover the dish with a watch glass.

Set up a preweighed 100 mL beaker and a long-stem funnel.

Use a large round filter paper to create a cone (demonstrated by your instructor) and place it in the funnel.
Holding the funnel over the beaker, transfer the extracted solution to the filter funnel and cover with the watch glass.

When the fluid has all drained through the filter paper, use your spatula to transfer the filter paper with the peanut meal to a labeled paper towel to dry. Leave the paper towel in the hood. Make sure your name is on your sample so that you can collect it later.

Place the 100 mL beaker containing the methylene chloride on a hotplate in the hood and evaporate the methylene chloride.

When the evaporation is complete, cool and re-weigh the beaker.

Part II: Analysis of the Peanut Oil

Using a transfer pipet, place 2 drops of the peanut oil from the beaker into a clean dry test tube.

Add 2 mL of ethanol to the test tube.

Add 1 drop of 2% potassium permanganate solution and swirl gently to mix the solution. Record your observations of changes to the solution over the next 5 min.

Set up a Bunsen burner.

Using a transfer pipette, place 3 drops of the peanut oil from the beaker into a clean, dry test tube. Add an equal amount of solid $KHSO_4$ to the test tube.

Heat the mixture vigorously in the flame, being careful not to point the test tube toward anyone.

Using your hand, waft the vapors being emitted from the test tube toward your nose to smell them. Record your observations.

Part III: Fehling's Test for Carbohydrates

Collect the now dry peanut meal from the hood.

Use your spatula to add a pea-sized amount to a clean, dry test tube.

Place 2 mL of hot water in the test tube and use your stirring rod to mix (smush) the meal into solution.
Add another 8 mL of water to the test tube and then split the suspension into two parts.

To one part, add 1 mL of concentrated HCl and place in a boiling water bath to hydrolyze for 10 min.

In a clean test tube, prepare Fehling's reagent;
    a. Mix 2.5 mL of solution with 2.5 mL of solution.
    b. Add 2 mL of this mixture to the test tube containing the peanut suspension that is not being heated.
    c. Record your observations. You may wish to leave the test tube for a few minutes to observe later.

Remove the other test tube from the water bath and neutralize it with 10% NaOH, testing with litmus paper to be sure that it is no longer acidic. Perform Fehling's test starting with step b above and record your observations.

Part IV: Xanthoproteic Test for Protein

Use your spatula to add a pea-sized amount of the peanut meal to a clean, dry test tube.
Place 5 mL of water in the test tube and use your stirring rod to mix (smush) the meal into solution.

Heat the test tube in a boiling water bath for 10 min.

Set up a beaker and filter as before.
Filter the suspension.

Place 1 mL of the filtrate in a clean, dry test tube and add 10 drops of concentrated nitric acid.

Heat the test tube carefully in a flame for a few minutes.
Cool the test tube and neutralize the filtrate with 10% NaOH.
Record your observations.

Report Contents and Questions

The purpose should include a majority of the topics and techniques covered in the experiment and the criteria for determining the success of the experiment. The procedure section should cite the lab manual and include any changes made to the written procedure during lab work.

The data section should include a table containing all of the observations you made during the experiment for each of the confirmation reactions. The mass data is the mass of the peanuts and the mass of oil recovered. The percent error is then calculated by subtracting the theoretical yield (the amount that should be acquired if the experiment had no human and experimental errors) from the actual yield (the amount that was actually obtain from the experiment) divided by the theoretical yield times a hundred. Make sure to record the raw values in your lab notebook, along with a sample calculation, and percent errors. Also include any observations made during the experiments.

The conclusion section should include several detailed paragraphs with the following information: **(a)** a discussion of the presence or absence of each biochemical constituent of the peanut and the test results that support these conclusions and **(b)** a discussion on the lipid content of your peanut. Finally, include all possible errors found in this experiment.

Answer the following questions:

1) You used a xanthoproteic test to confirm the presence of proteins in this lab. What is another test that you could use to confirm the presence of proteins? How does it work?

2) What does acrolein smell like? When have you smelled something similar to acrolein? What (chemically) caused the smell?

Experiment 29
Laboratory Preparation

Name: _____ Date: _____

Instructor: _____ Sec. #: _____

Show all work for full credit.

1) Read the background, procedure, and report sections of the lab experiment carefully and develop a hypothesis of what information you expect to gain from the completion of the lab experiment.

2) Create any and all tables you might need to collect data during the experiment and then transfer those tables into your lab manual for use during the lab.

Experiment 29
Prelaboratory Assignment

Name: _____ Date: _____

Instructor: _____ Sec. #: _____

Show all work for full credit.

1) What is a nonreducing sugar? How does Fehling's Test work to show the presence of non reducing sugars?

2) There is a general belief that the tryptophan found in turkey is what causes the sleepiness that often follows the Thanksgiving meal. Act as a myth buster and investigate the accuracy of this belief. Is there a lot of tryptophan in turkey? How does tryptophan cause sleepiness? Or does it?

3) Hydrolysis is a process the human body uses to break down food in the process of digestion. Look up and explain the process of three enzymes that use hydrolysis as part of their enzymatic function.

Appendix A
Glossary of Terms Used in This Manual

Absorbance (A): a logarithmic function of the percent transmission of a wavelength of light through a liquid.

Absorbent: anything that absorbs.

Aliquot: a sample that is a definite fraction of the whole.

Barometer: an instrument that measures the pressure of the atmosphere.

Buret: an apparatus for delivering measured quantities of liquid or for measuring the quantity of liquid or gas received or discharged. It consists essentially of a graduated glass tube, usually furnished with a small aperture and stopcock.

Capillary action: the movement of liquid within a material against gravity as a result of surface tension.

Chromatography: techniques for separating molecules based on differential absorption and elution. Also a term for separation methods involving flow of a fluid carrier (mobile phase) over a stationary absorbing phase.

Cuvette: a transparent or translucent box-shaped container with precisely measured dimensions for holding liquid samples to be put into a spectrophotometer.

Dry ice: solidified carbon dioxide.

End point of a titration: that point at which the indicator shows a change.

Error of the titration: the difference in volume of the titrant used between the equivalence point and end point.

Extinction coefficient: a measure of how strongly a chemical species absorbs light at a specific wavelength. Also called the molar absorptivity or molar extinction coefficient (ε) of a chemical species at a given wavelength, it is an intrinsic property of the chemical species. The actual absorbance of a sample is also dependent on its thickness (L) and the concentration (c) of the species

Hard water: water containing high concentrations (2 to 4 mmol/L) of soluble salts of calcium and magnesium, and sometimes iron.

Indicator: a chemical compound that exhibits an observable change near the equivalence point due to a change in oxidation–reduction or pH.

Meniscus: the curved surface of the liquid at the open end of a column.

Mobile phase: the phase that moves along the stationary phase. It is the solvent in paper chromatography and thin-layer chromatography.

Molar absorptivity (ε): *see* extinction coefficient.

pH range for an acid–base indicator: the pH interval over which the indicator shows a change in color.

Primary standard: a highly purified stable chemical compound of known chemical composition that can be weighed with great accuracy.

Qualitative: term used to describe observations that do not involve measurements and numbers.

Quantitative: term used to describe observations that involve measurements and numbers.

Retention factor (R_F) : in chromatography, the distance traveled by the compound divided by the distance traveled by the solvent.

Separatory funnel: a piece of laboratory glassware used in liquid term used to describe liquid extractions to separate (partition) the components of a mixture between two immiscible solvent phases of different densities.

Serial dilution: a series of dilutions that amplifies the dilution factor quickly, beginning with a small initial quantity of material. The source of dilution material for each step comes from the diluted material of the previous step.

Spectroscope: an optical instrument used to produce spectral lines and measure their wavelengths and intensities; used in spectral analysis.

Standard cell: an electrochemical cell whose half-cells are both in the standard state of 25 ºC and 1.0 atm, and are immersed in electrolytes at a concentration of 1.0 M.

Standardization of a solution: the process of finding the exact concentration of a solution.

Stationary phase: in chromatography, the non mobile phase contained in the chromatographic bed.

Titrant: a titration reagent added from the buret to the sample solution. When referring to a specific titration, the titrant is placed second.

Transmittance (%T) : the fraction of radiant energy that passes through a substance at a given wavelength.

Vernier scale: a small movable scale that slides along a main scale; the small scale is calibrated to indicate fractional divisions of the main scale.

Volumetric method: a method in which the analysis is completed by measuring the volume of a solution of established composition necessary to react completely with the unknown substance.

Appendix B
Excel Tutorial

The competent use of a spreadsheet program such as Excel is an important skill for a scientist to master. In addition to simplifying repetitive computations, it allows the user to produce publication-quality graphics. In science more so than any other area, a picture can indeed be worth a thousand words, but only if it is accurately created. What follows is a very brief summary of how to use Excel to calculate some simple algebraic formulas and common statistical values, and to create graphs from typical experimental data.

Getting Started

When you open Excel, you will see a spreadsheet with alphabetically labeled columns and numerically labeled rows. This means that each cell in the spreadsheet has a unique alphanumeric label. It is these labels that can be input into formulas.

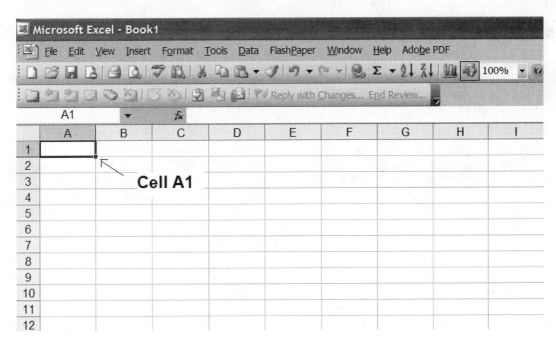

Note: Your screen may vary a bit from the one shown here. Commands not shown on your toolbar can usually be accessed from one of the menus. If you will be using a particular command frequently, the button can be added to your toolbar.

You can enter data into each cell, in either alphabetic or numeric form, by simply typing whatever you need into each cell. You can move from cell to cell using the Enter (down one cell) or Tab (across one cell) key after you type in your entry. The number of decimal places displayed in the data can be changed by highlighting the data, selecting Cells under the format menu, and choosing the Number tab. the new menu shown, select Number, and you will see a display that allows you to select the number, of decimal places shown. **This is especially useful for displaying calculation results with the correct number of significant figures**.

Basic Math

Once you have your data in the columns as you want it, you need to be able to use simple expressions to manipulate the data. A mathematical equation in Excel is always preceded by an equal sign, =.

For basic addition, subtraction, multiplication, and division of the data in the cells, the formulas are:

Operation	Example Input
Addition	=A1+B2
Subtraction	=A1-B2
Multiplication	=A1*B2
Division	=A1/B2

Note: Combinations of these operations and more complex expressions can be entered using parentheses between operations.

For example, $V = \dfrac{nRT}{P} = \dfrac{1(0.08206 \frac{\text{L·atm}}{\text{mol·K}})(273\,\text{K})}{1\,\text{atm}} = 22.4\,\text{L}$, would be calculated as shown below:

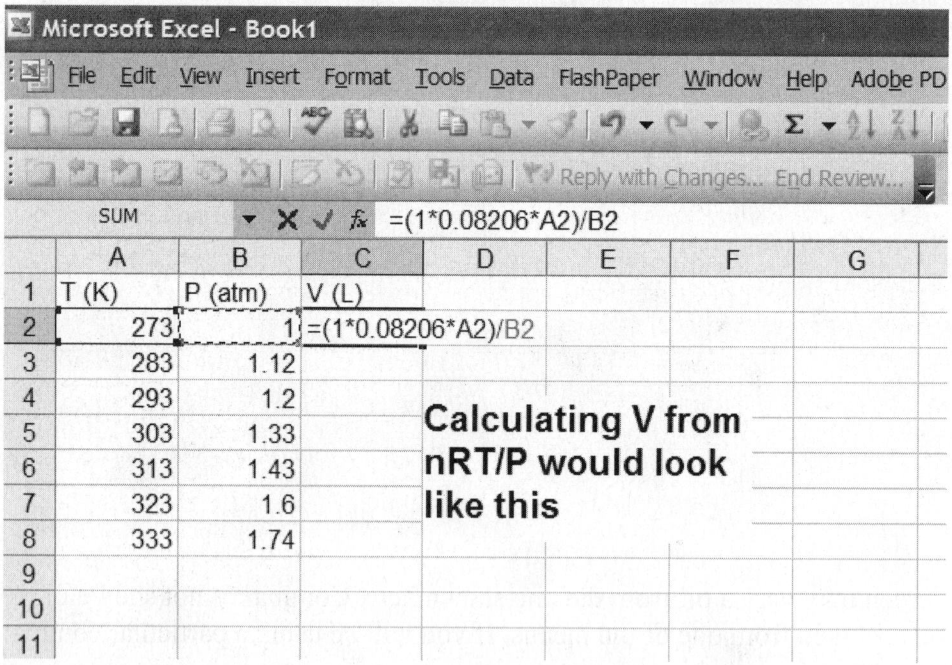

Now, if you want to continue with the calculation but don't want to type the equation in several times, you can copy the equation simply by right-clicking your mouse at the bottom right of the cell you want to copy, dragging it the length of the column you want it to fill, releasing the mouse, and hitting Enter. You should notice that the expressions remain the same, but the cell values change running down the column as the referenced cells change in value. (Alternatively, you can use the mouse to select the cell you want to copy, highlight the column of cells where you want the formula, select Fill from the Edit menu, and choose the option you want.)

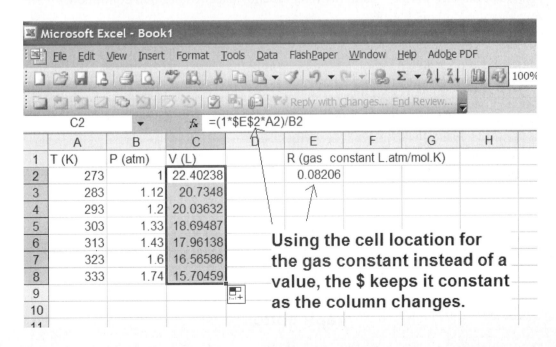

If you want a value to remain constant and not change as you copy values down a column, this can be done by simply placing a $ in front of the number and letter of the cell containing the value you do not want to change. For example, if cell E2 contains a constant in your formula, enter it as E2. Notice that the cell contains only the actual number. The identifying words are in other cells. Excel will not calculate non-numeric values.

Graphing

To create a graph in Excel, first input the data you want to graph with the X values in the column on the left and the Y values in the column to its right.

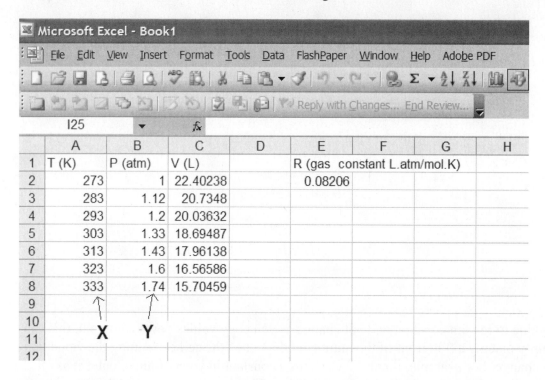

Highlight the columns you wish to graph, with headings, and press the graph button on the toolbar. If the graph button is not on your toolbar, go to the Insert menu, select Chart, and continue with the instructions below.

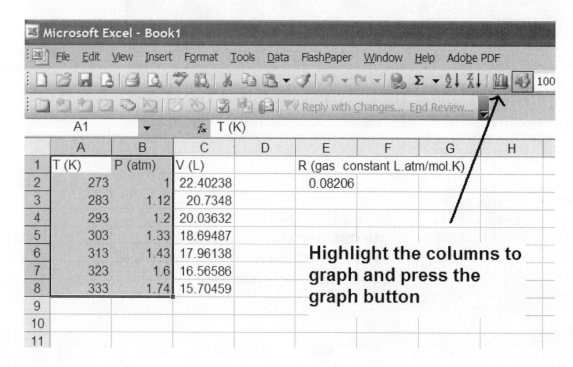

Select the type of graph you want to make. Most graphs for the laboratories in this manual will be XY scatter plots.

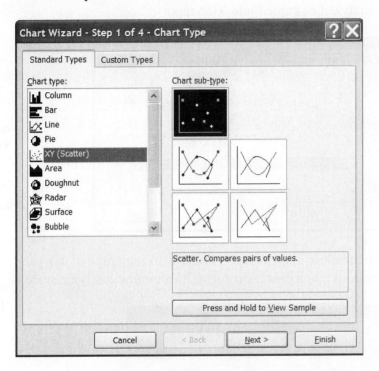

Once you select the type of graph, you will then be asked to complete the parameters for the graph. Select Next from the screen for Step 2. Begin Step 3 with the input of titles for the *x*- and *y*-axes and the overall title for the graph. Then you can use the tabs to add or remove gridlines, set the placement of the legend, or choose to label the individual data points.

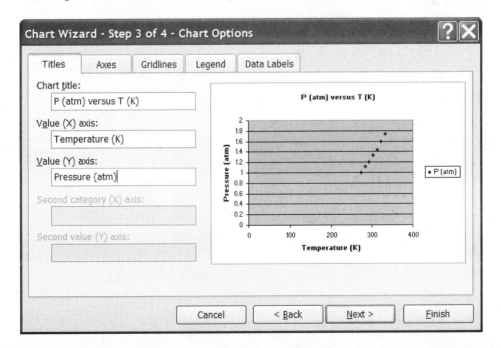

After setting the graph up the way you would like it to appear, hit Next and choose to have the graph either be its own page or appear as a graphic within the spreadsheet. Then click Finish. You will have a completed graph ready to be analyzed or copied into a lab report.

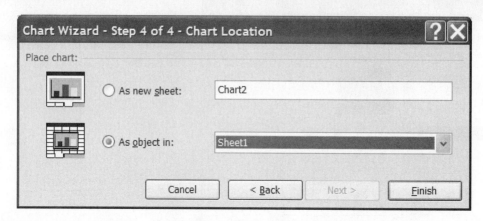

Additional changes to the graph as described below must be made to the working graph that is part of the spreadsheet. Once the graph is copied into your final document, it cannot be easily changed.

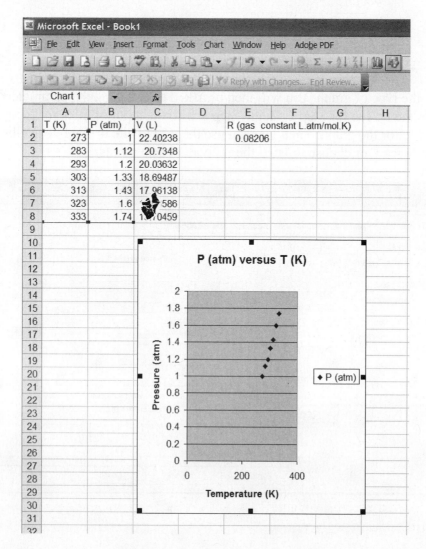

To adjust the scale so that your data points fill the graph completely, double-click on the axis you want to adjust and change the maximum or minimum value as needed. If you plan to use your graph to predict a value of X or Y, be sure your scale goes a bit beyond that value.

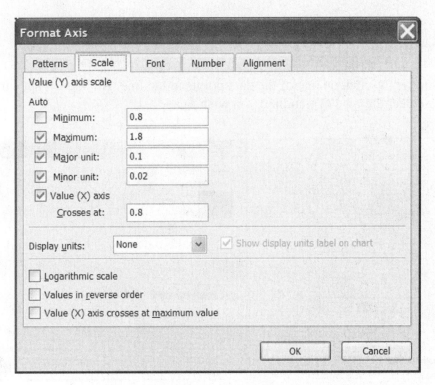

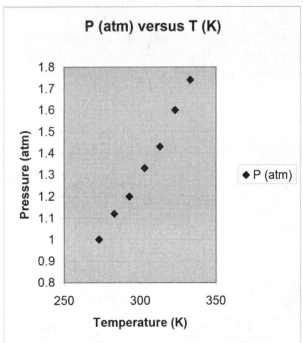

Notice that by adjusting the scale of both the *x*- and *y*-axes, the graph is now much larger and centered, making it easier to analyze.

The first piece of analysis you will want to complete with respect to your graph is the addition of a trendline, a best-fit line that follows the tendency of the data and can express that tendency in an equation. For a relatively straight line like the one in our example graph, a linear trendline is the best choice.

Adding a Trendline

To add a trendline, right-click on one of the data points in the line. A box will appear. Select Add Trendline. Then select the type of trendline you wish to add.

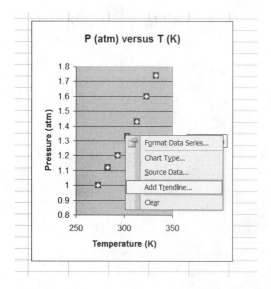
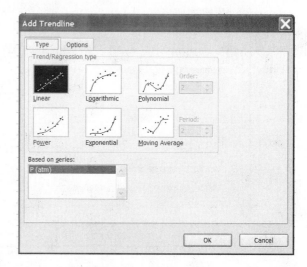

Next, click on the Options tab. Here, you can choose to display the slope and equation of the line to your graph. For some data sets, you will predict values using the forecast option. The R-squared value is a statistical measure of how well your data fits the $y = mx + b$ equation; a value very close to 1 indicates a good fit. This value is often displayed to verify the credibility of the data and the graph.

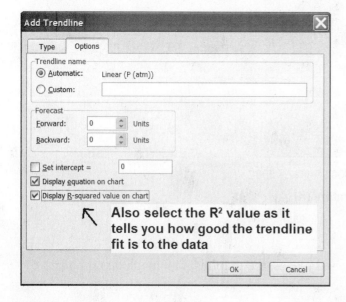
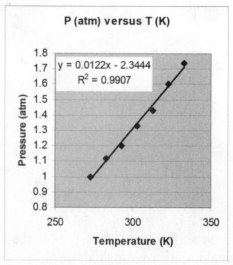

Add Error Bars

Right-click on a data point on the line and select Format Data Series. On the Y-Error Bars tab, select Both Display. Here, you have a choice of the type of error bars to display on the graph. Select Custom and then click on each selection icon.

In Excel 2007, the Error Bars functions can be found in the Chart Tools: Layout area. Select Error Bars and follow the directions below.

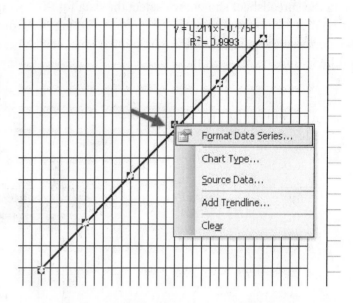

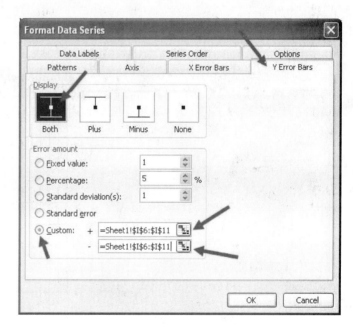

Using the spreadsheet shown, we select the data for 90% Confidence Interval (cells I6:I11) to display this confidence interval as the error range. Selecting the same range for both the + and – bars will give equal error above and below the data point. (This means there is a 90% probability that the "true" value will lie within the range of these bars).

Next, right-click on one of the error bars and select Format Error Bars.

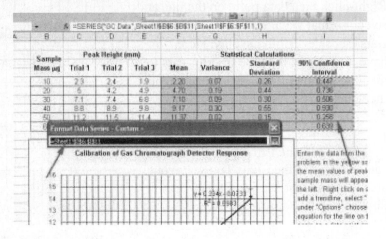

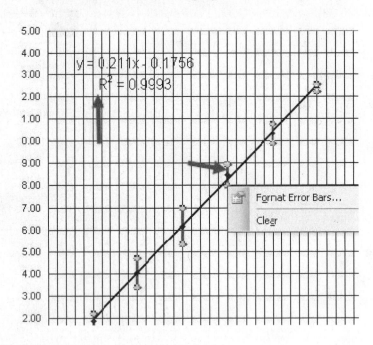

Calibration of Gas Chromatograph Detector Response

On the Patterns tab, and check Custom. You can change the color of the error bars if you wish and the thickness of the line so that they show up better on the graph.

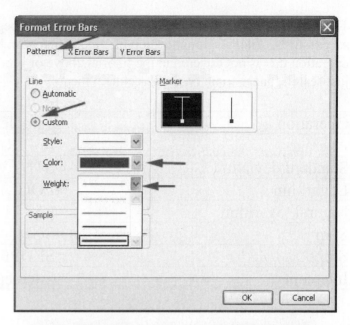

Spreadsheets can be used for many other calculations; a few common ones are listed in the following table.

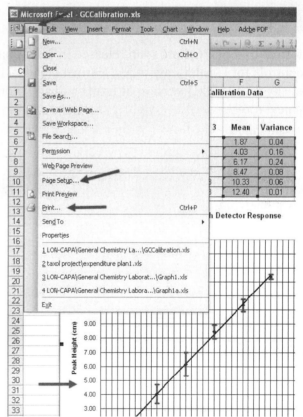

The information thus far should allow the students to begin the lab processes and experiments successfully. Another thing to keep in mind is that Excel has a good help section. If you need an operation that is not covered in the tutorial above or you need more details than are provided here, just search for the topic on the Excel help menu.

Operation	Example Input
Average	=AVERAGE(A1:A11)
Standard deviation	=STDEV(A1:A11)
Logarithm	=LOG(A1)
Natural logarithm	=LN(A1)
Sum	=SUM(A1:A11)
Slope	=SLOPE(A1:A11)
Intercept	=Intercept(A1:A11)

Excel 2007

The format of the Excel program released in 2007 changed the interface for the graphing protocols. The controls for Design Layout and Format are all found on the main toolbar. The directions given previously for the 2003 version of Excel will still work once you have opened the correct section of the new toolbar.

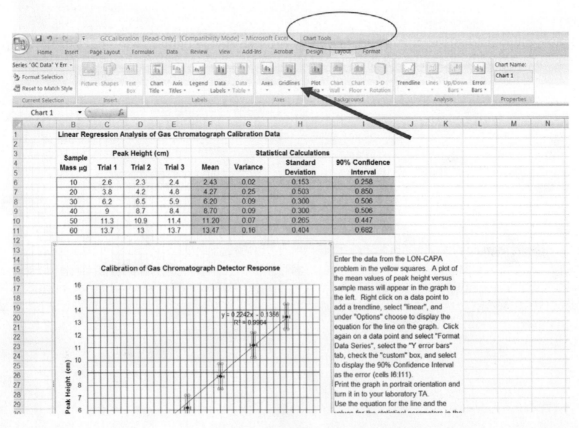

Appendix C
Volumetric Glassware

There are three principal types of volumetric glassware: *pipets*, *burets*, and *volumetric flasks*.

Markings on Volumetric Glassware

Volumetric glassware is designed to contain a certain volume or to deliver a certain volume of reagent. Volumetric equipment you encounter may be marked either TC (to contain) or TD (to deliver). The temperature at which the equipment is to be used is also often printed on the equipment. When choosing a piece of equipment to use, note how it is marked.

Cleaning Volumetric Glassware

All glassware should be washed with a soap solution and rinsed three times with tap water. If your local water supply is quite hard, also rinse glassware three times with deionized water from your wash bottle. Always fill your wash bottle with deionized water from the special taps in the lab, NEVER with regular tap water. Also note: DO NOT USE THE EYE WASH AS A SINK.

Ordinary standards of cleanliness are not sufficiently stringent for pipets, burets, and volumetric flasks. Solutions should drain in an even film on the inside surfaces of these vessels. Slight traces of oil coating cause aqueous solutions to break up into drops, rendering volume measurements inaccurate. If solutions do not drain evenly on the inside surfaces of the calibrated glassware, consult your lab instructor. Once your glassware is clean, immediate and thorough rinsing according to the directions above will generally preserve its cleanliness. The glassware should be allowed to air dry or wiped dry with a Kimwipe. Paper towels are inappropriate for drying chemical glassware because they shed fibers that can contaminate your experiment.

After using a chemical solution in a buret, fill it with tap water, run some of the rinse through the tip, and then rinse thoroughly with tap, and then deionized, water, three times each. Finally, always rinse the buret with a small volume of the new titrant prior to beginning the titration.

Reading the Meniscus

A liquid confined in a narrow tube exhibits a marked curvature of the surface called the *meniscus*. Common practice is to read the bottom of the meniscus when using volumetric glassware; with opaque solutions such as $KMnO_4$, it is necessary to read the top. It is often advantageous to highlight the meniscus by placing a white card with a heavy dark line upon it behind the container, with the top of the dark band a few millimeters below the meniscus. Light reflected from the liquid surface makes the meniscus stand out effectively. The eye of the observer should be horizontal to the meniscus. Noting the nearest calibration line that goes completely around the container, the observer can ascertain if the eye is at the correct level.

The Pipet

Solutions are drawn into the pipet with rubber bulbs or dial-up pipetters. (It is important to keep the inside of the bulb dry; that is, it must not become contaminated. Be careful not to suck up the liquid into the bulb.) Standardized technique to fill and drain a pipet is necessary to minimize error.

First, if it is not dry, the pipet is rinsed with two or three small portions of sample so that the entire inner wall is wetted with solution before pipetting begins. The outside of the tip of the pipet should be kept free of drops of solution, as there is the danger that drops might be transferred to the receiver.

The bulb is then put on the pipet top as loosely as possible. The solution is drawn to a point at least an inch above the calibration line, the rubber bulb or pipetter is withdrawn, and the *index* finger (not the thumb) is placed over the opening. The outside of the long delivery tube is wiped off with a Kimwipe®, and the solution is allowed to drain down to the mark by varying the angle of the index finger. Both the finger and the pipet opening should be dry for best results.

The solution is allowed to flow freely but without spattering into the receiver. Wait 5 sec after the flow has ceased for the pipet to drain, then touch the tip of the pipet to the moist side of the receiver. A small residue will remain in the pipet. Do not blow this out, as this will not allow for a consistent measurement of liquid from experiment to experiment. Follow this procedure rigorously, and measured volumes with the precision of *one part per thousand* (1 ppt) will be delivered.

The Mohr Pipet

The Mohr pipet is a hybrid between a buret and a pipet. It is a pipet that has the calibrations of a buret but not a stopcock. The same operating and cleaning procedures described for the pipet and buret apply.

The Buret

Burets provide a means for accurately delivering a volume of liquid; they consist of a calibrated tube and stopcock arrangement that allows a flow from the tip to be controlled. Scrupulous cleanliness is mandatory. Recall that the buret is to be rinsed with the solution it is to contain and filled all the way to the tip. Also note that the stopcock should be liquid tight so that it does not leak! During the titration demonstration by your lab instructor, pay particular attention to the way he or she checks the buret tip just below the stopcock for air bubbles. After the solution has been delivered, note that the buret must be given about 30 sec to drain before a reading is taken.

Teflon stopcocks are made of very chemically inert materials, but a few rules for care are in order:

1) NEVER USE ABRASIVE MATERIALS to clean either the stopcock or barrel. This includes brushes.

2) A Teflon washer is generally placed adjacent to the end of the barrel so that minimal friction is created upon turning the stopcock. These washers are small, so be careful not to lose yours when you are cleaning the stopcock.

3) Teflon stopcocks can be easily scored around the bore if they are rotated when solid particles are lodged between the plug and barrel. Once scored, the stopcock may leak.

4) If Teflon stopcocks are used with liquids that are corrosive to glass, such as alkalis, *rinse the stopcock thoroughly with water after use*. Do not allow the liquid to evaporate. The remaining concentrated solution will attack the glass surface, and the eventual solids may also mar the Teflon surface if the stopcock is then rotated.

5) When it is not in use, store the buret in a dust-free space with the stopcock loose in the barrel. Although Teflon is tough and unbreakable, it is softer than glass and has a tendency to conform to the glass surface, including eventual expansion into the ports of the barrel.

The Volumetric Flask

The principal use of the volumetric flask is to prepare solutions of a specified concentration, either in preparation of standard solutions or in dilution of samples of known volumes prior to taking aliquot portions with a pipet. The flask has been calibrated to reflect a given volume and temperature as marked on the flask.

After transferring the solute, fill the flask about half-full and swirl to mix. Add more solvent and mix again. Bring the liquid level almost to the mark, allow time for drainage down the neck of the flask, then *use a medicine dropper* to make the necessary additions of solvent. Firmly stopper the flask and invert repeatedly to ensure uniform mixing. Loosen the stopper to allow the solution to drain back into flask before removing the stopper completely. The contents *must* be homogeneous to achieve valid results.

NOTE: Solutions added should be at room temperature for accurate volume calibration. A 60 °C, solution suffers a considerable volume loss on cooling to room temperature. Since the flask is often made from soft glass, alkali solutions should not be stored in them for indefinite periods of time.

Preparing a standard solution often requires that a known weight of solid solute be introduced into a volumetric flask. This can be accomplished in two ways:
1) Insert a powder funnel into the neck of the flask and add the solid sample. After transfer, the solid is washed off the funnel into the flask with deionized water or the appropriate solvent, and the resulting solution is diluted to the correct volume.

2) Dissolve the sample in a *clean* beaker or flask with a small portion of water and then transfer the solution into the volumetric flask. The beaker is then rinsed three times with small amounts of water, and each rinse is added to the volumetric flask. The advantage of using this procedure is that solids that do not easily dissolve in room-temperature water can be heated in the beaker and then transferred to the volumetric flask. Be sure the solution is at room temperature before diluting to the final volume.

Appendix D:
Standard Reduction Potentials (in Volts), 25 °C

Reaction	$E°$
$F_2 + 2e^- \rightarrow 2F^-$	+2.87
$Co^{3+} + e^- \rightarrow Co^{2+}$	+1.81
$PbO_2 + 4H^+ + SO_4^{2-} + 2e^- \rightarrow PbSO_4(s) + 2H_2O$	+1.69
$MnO_4^- + 8H^+ + 5e^- \rightarrow Mn^{2+} + 4H_2O$	+1.51
$PbO_2 + 4H^+ + 2e^- \rightarrow Pb^{2+} + 2H_2O$	+1.46
$Cl_2 + 2e^- \rightarrow 2Cl^-$	+1.36
$Cr_2O_7^{2-} + 14H^+ + 6e^- \rightarrow 2Cr^{3+} + 7H_2O$	+1.33
$O_2 + 4H^+ + 4e^- \rightarrow 2H_2O$	+1.23
$Br_2 + 2e^- \rightarrow 2Br^-$	+1.09
$NO_3^- + 4H^+ + 3e^- \rightarrow NO + 2H_2O$	+0.96
$Hg^{2+} + 2e^- \rightarrow Hg$	+0.85
$Ag^+ + e^- \rightarrow Ag$	+0.80
$Fe^{3+} + e^- \rightarrow Fe^{2+}$	+0.77
$I_2 + 2e^- \rightarrow 2I^-$	+0.54
$Cu^+ + e^- \rightarrow Cu$	+0.52
$Fe(CN)_6^{3-} + e^- \rightarrow Fe(CN)_6^{4-}$	+0.36
$Cu^{2+} + 2e^- \rightarrow Cu$	+0.34
$Cu^{2+} + e^- \rightarrow Cu^+$	+0.15
$Sn^{4+} + 2e^- \rightarrow Sn^{2+}$	+0.15
$2H^+ + 2e^- \rightarrow H_2$	0.00
$Fe^{3+} + 3e^- \rightarrow Fe$	−0.04
$Pb^{2+} + 2e^- \rightarrow Pb$	−0.13
$Sn^{2+} + 2e^- \rightarrow Sn$	−0.14
$Ni^{2+} + 2e^- \rightarrow Ni$	−0.26
$Co^{2+} + 2e^- \rightarrow Co$	−0.28
$PbSO_4 + 2e^- \rightarrow Pb + SO_4^{2-}$	−0.359
$PbI_2 + 2e^- \rightarrow Pb + 2I^-$	−0.365
$Cr^{3+} + e^- \rightarrow Cr^{2+}$	−0.40
$Cd^{2+} + 2e^- \rightarrow Cd$	−0.40
$Fe^{2+} + 2e^- \rightarrow Fe$	−0.45
$Cr^{3+} + 3e^- \rightarrow Cr$	−0.74
$Zn^{2+} + 2e^- \rightarrow Zn$	−0.76
$2H_2O + 2e^- \rightarrow H_2(g) + 2OH^-$	−0.83
$V^{2+} + 2e^- \rightarrow V$	−1.18
$Mn^{2+} + 2e^- \rightarrow Mn$	−1.18
$Al^{3+} + 3e^- \rightarrow Al$	−1.66
$Mg^{2+} + 2e^- \rightarrow Mg$	−2.37